Reiman Gardens

A Guide To Lawn, Garden & Home Pest Control Products

by
W.T. THOMSON

THOMSON PUBLICATIONS
P.O. BOX 9335
FRESNO, CA 93791
(209) 435-2163
FAX (209) 435-8319

**Copyright 1996 by Thomson Publications
Printed in the United States of America
ISBN – Number 0-913702-62-5**

All rights reserved. No part of this book may be reproduced in any form by photostat, microfilm or any other means, without written permission from the author.

THIS BOOK FOR REFERENCE ONLY

READ THE LABEL CAREFULLY

HOW TO USE THIS BOOK

This is a convenient quick guide to the products and suppliers of the lawn and garden industry's products in the U.S. It is designed to be used by nursery store personnel and homeowners as a problem solving guide to homeowner pest control problems. The book is divided into four parts. First, the manufacturer and the products they supply. Section II contains the various chemical products, their active ingredient and what pests they control. Section III contains the various usages that appear on the labels and what labels these usages appear on. Section IV is a cross reference of scientific and common names of common garden plants.

Since this is our first attempt at a book of this type the author welcomes any suggestions to make the book more user friendly with future editions.

Section I – Manufacturers & Their Products – This section lists the major suppliers of the U.S. lawn and garden pest control industry. The companies are listed with their product lines broken down into insecticides, herbicides, fungicides and miscellaneous chemicals. Most of the information is from actual labels, but some companies only submitted catalogs so they could not be included in the rest of the text.

In the U.S. each pesticide label must be registered in each state into which it is sold. Therefore, many of these products are only sold on a regional basis and may not be readily available in your particular area. A call to the individual company, as to availability, would be advised if you can not find a paricular product.

Section II – Products & What They Control – All pesticide products have a generic name with various trade names. Sometimes they are the same (i.e. Diazinon). In this section the generic name is listed followed by the products sold that contain that active ingredient. Also, the % active ingredient is listed for the individual products sold. The pests that the activie ingredient controls are then listed. Most of the labels do not list all of the pests they control. This is based on decisions made by each individual supplier. Therefore, you should read the label completely before using.

Note: *For simplicity, we have listed products that contain both an insecticide and fungicide in the fungicide section only. Therefore, throughout the book look for these combination products in the fungicide section.*

Section III – Useage Of Products – This section lists the individual plants, vegetables, trees, shrubs, etc. and which products refer to the usage on them. If a particular plant is listed on the label, the labels containing that usage are listed, again broken down into insecticides, herbicides and fungicides. Many labels are very general (i.e., use on flowering plants) so they would be listed under flowering plants (general) rather than specific flowering plants. Again, before selecting a product be sure the usage appears on the label.

Section IV – Plant Names – Many plants are recognized both by the scientific name and the common name. Often pesticide labels use one or the other and the consumer is confused if both names are not known. An attempt to cross reference the names of common homeowner usage plants have been made by listing the scientific name with the common name next to it, then the common name with the scientific name next to that. If a plant's common and scientific name are the same it was not listed.

This book is only a guide to the products that can be used by the homeowner in pest control situations. There are many regional suppliers of these products that are not listed, although their products are on the shelf at your local nursery. In future editions we will attempt to list as many of these suppliers as possible.

Also, please remember that suppliers are continuing to add and delete products from their lines and change labels as new usages appear. Some of the products listed will not be available in the future and all of the suppliers will continue to add new products and change lables Since this book is a guide only to making the right decision on what product to use, remember to read the label carefully and follow its directions.

– **W.T. Thomson**

A Guide To Lawn, Garden & Home Pest Control Products

TABLE OF CONTENTS

Section I	**Manufacturers & Their Products**	1
Section II	**Products & What They Control**	19
	Insecticides	20
	Herbicides	27
	Fungicides	32
	Miscellaneous Chemicals	36
Section III	**Useage Of Products**	39
	Insecticides	40
	Herbicides	72
	Fungicides	90
Section IV	**Plant Names**	105
	Scientific Names (Common Name Cross Reference)	106
	Common Names (Scientific Name Cross Reference)	114

A Guide To
Lawn, Garden & Home
Pest Control Products

Section I   **Manufacturers & Their Products**

MANUFACTURERS & THEIR PRODUCTS

AGR EVO ENVIRONMENTAL HEALTH

95 Chestnut Ridge Rd.
Montvale, NJ 07645
(201) 307-9700
FAX (201) 307-3281

Insecticides

50% Malathion
Alter (methoprene/permethrin)
Alter Flea Control EC (methoprene)
Ant Killer Granules (chlorpyrifos)
Diazinon 4E
Diazinon 5% Granules
Diazinon 25%
Dursban Lawn & Perimeter Granules (chlorpyrifos)
Dursban (chlorpyrifos)
Fire Ant Granules (chlorpyrifos)
Flea, Tick & Mange Dip (chlorpyrifos)
Intercept H&G Grub Control Granules (bendiocarb)
Intercept Insect Control (resmethrin/permethrin)
Intercept Insect Control - Vegetable Lawn & Garden Spray (permethrin)
Primicide Fire Ant Dust (chlorpyrifos)
Sevin 5% Dust (carbaryl)
Sevin 10% Dust (carbaryl)
Termi-Chlor (chlorpyrifos)
Ultra S.S. C. (resmethrin/chlorpyrifos)
Vikor (cypermethrin)
Vikor Dual Action Home & Patio Spray (cypermethrin)
Vikor Ready to Use Home Pest Control (tralomethrin)

Herbicides

Finale Concentrate (glufosinate-ammonium)
Finale Ready to Use (glufosinate-ammonium)
Finale Super Concentrate (glufosinate-ammonium)
Weed Warrior Dandelion & Broadleaf Killer (2,4-D/MCPP/2,4-DP)
Weed Warrior Spot Weed Killer (2,4-D/MCPP/2,4-DP)

Fungicides

Procide (tridimefon)
Procide G Systemic Lawn Fungicide (tridimefon)
Wettable Dusting Sulfur

BLACK LEAF PRODUCTS

Rigo Company (Wilbur Ellis)
P.O. Box 189
Buckner, KY 40010
(502) 222-1466

Insecticides

5% Diazinon Dust
5% Diazinon Granules
5% Sevin Garden Dust (carbaryl)
25% Diazinon
50% Malathion Spray
Ant Killer Powder (diazinon)
Carpenter Ant Killer (chlorpyrifos)
Cygon 2-E Soil Drench (dimethoate)
Dormant Spray (petroleum oil)
Dursban (chlorpyrifos)
Dursban Lawn Insect Control (chlorpyrifos)
Fire Ant Killer (diazinon)
Liquid Sevin (carbaryl)
Mosquito Spray Concentrate (malathion/methoxychlor)
Rose & Flower Insect Killer II (pyrethrin)
Rose Guard (disulfoton/trifluralin)
Sevin 50% WP (carbaryl)
Termite Killer (chlorpyrifos)
Thuricide (B.t.)

Herbicides

Crabgrass Killer (MSMA)
Eptam Weed Control (EPTC)
Grass Weed & Vegetation Killer (sodium chlorate)
Lawn Edging Liquid (sodium chlorate/diquat)
Lawn Weed Killer (2,4-D/2,4-DP/MCPP)
Spot Weed Killer (2,4-D)
Weed & Grass Killer (sodium chlorate/diquat)

Fungicides

Bordeaux (basic copper sulfate)
Lawn & Garden Fungicide (chlorothalonil)
Liquid Copper Fungicide (copper-fatty & rosin acids)
Rose & Ornamental Fungicide (thiophanate-methyl)

BONIDE PRODUCTS INC.

2 Wurz Road
Yorkville, NY 13495
(315) 736-8231
FAX (315) 736-7582

Insecticides

Ant Dust (diazinon)
Bacillus Thuringiensis
Borer Miner Killer 5 (lindane)
Carpenter Ant Control (chlorpyrifos)
Colorado Potato Beetle Beater (B.t. var San Diego)
Cygon System Insecticide (dimethoate)
Diazinon 12 1/2% E
Diazinon Insect Control
Flea Beater (bendiocarb)
Garden Soil Insecticide Diazinon 5% G
Home Pest Control Concentrate (chlorpyrifos)
Home Pest Control RTU (chlorpyrifos)
Horticultural & Dormant Spray Oil (petroleum oil)
Imidan (phosmet)
Insecticidal Soap Concentrate (fatty acids)
Insecticidal Soap for Fruit & Vegetable (fatty acids)
Japanese Beetle Killer (pyrethrin)
Kelthane Liquid (dicofol)
Liquid Sevin (carbaryl)
Malathion 50% EC
Methoxychlor 25% E
Mite & Insect Spray
   (malathion/lindane/carbaryl/dicofol)
Mite Beater (permethrin)
Mole Cricket Killer (diazinon)
Mosquito Larvacide (mineral oil)
Organic Greenhouse, House & Vegetable Spray
   (pyrethrin)
Permethrin Turf & Ornamental
Pest Control Concentrate (permethrin)
Rose & Floral Spray Bomb (tetramethrin)
Rose Flower & Ornamental Insect Spray (pyrethrin)
Sevin 5% Dust (carbaryl)
Sevin 10% (carbaryl)
Systemic Granules (disulfoton)
Systemic Rose & Flower Care (disulfoton)
Termite & Carpenter Ant Control (chlorpyrifos)
Tomato & Vegetable Ready to Use (pyrethrin)
Tomato Pepper Vegetable Spray (pyrethrin)

Herbicides

Brushkil (2,4-D/dicamba)
Crabgrass Preventer & Weed Killer (siduron)
Eptam 2.3 (EPTC)
Garden Turf & Ornamental Herbicide 5G (DCPA)
Grass & Weed Killer (diquat)
Lawn Spot Weeder (2,4-D/MCPP/dicamba)
Lawn Weed Killer (2,4-D/dicamba)
MSMA Crabgrass Killer
Poison Ivy & Brush Killer (2,4-D/2,4-DP)
Poison Oak & Ivy Killer (diquat)
Turf & Ornamental Herbicide 75% WP (DCPA)
Weed & Feed Granules 17-4-4
   (2,4-D/MCPP/dicamba)
Weed & Feed Liquid 15-2-3 (2,4-D/MCPP/dicamba)
Weed & Grass Killer (glyphosate)

Fungicides

Bayleton Systemic Fungicide (triadimefon)
Bulb Dust (thiram/methoxychlor)
Captan 50% WP
Complete Fruit Tree Spray
   (captan/methoxychlor/malathion/carbaryl)
Copper Spray or Dust (basic copper sulfate)
Fruit Tree Spray
   (methoxychlor/captan/malathion/carbaryl)
Liquid Sulfur
Mancozeb Plant Fungicide
Rose & Flower Spray or Dust
   (methoxychlor/captan/malathion/carbaryl)
Sulfur Plant Fungicide
Tomato Potato Dust (maneb/carbaryl)

Miscellaneous

Berry & Fruit Set Spray (cytokinin)
Mole Tox II (zinc phosphide)
Mosquito Beater (napthalene)
Rabbit-Deer Repellent (thiram)
Rooting Powder (B.t.)
Slug Snail & Sowbug Bait (carbaryl/metaldehyde)
Slug-N-Snail Beater (metaldehyde)
Tomato & Blossom Set Spray (cytokinin)

DEXOL INDUSTRIES

1450 West 228th Street
Torrance, CA 90501
(310) 326-8373
Fax (310) 325-0120

Insecticides

Ant Killer Dust II (chlorpyrifos)
Ant Killer Granules (chlorpyrifos)
Ant, Roach & Spider Control (diazinon)
Ant, Roach & Spider Killer II (pyrethrin/permethrin)
Aphid, Mite & Whitefly Killer II (resmethrin)
Carpenter Ant Killer (chlorpyrifos)
Diazinon 2% Granules
Diazinon 5% Granules
Diazinon 25% Spray
Dipel Insect Control (B.t.)
Dormant & Summer Oil Spray (petroleum oil)
Dursban Granules Insect Control (chlorpyrifos)
Dursban Lawn Insect Killer (chlorpyrifos)
Dursban Lawn Insect Spray (chlorpyrifos)
Fire Ant Granules II (chlorpyrifos)
Flea-Free Carpet Treatment (pyrethrin)
Flea & Tick Spray (chlorpyrifos)
Flying Insect Killer (tetramethrin/sumithrin)
Home Pest Control Concentrate (chlorpyrifos)
Hornet & Wasp Killer II (tetramethrin/sumithrin)
Household Insect Control (pyrethrin)
Malathion Insect Control
Predator Carpenter Ant Killer Dust (bendiocarb)
Predator Home Insect Killer II (chlorpyrifos)
Predator Roach Powder (boric acid)
Predator Termite Killer Dust (bendiocarb)
Rose & Floral Insect Control (pyrethrin)
Rose & Floral Insect Killer (pyrethrin)
Sevin Brand 5% Dust (carbaryl)
Sevin Liquid Insect Killer II (carbaryl)
Systemic Rose Care (disulfoton)
Systemic Shrub & Flower (dimethoate)
Tender Leaf Aphid, Mite & Whitefly Killer
 (tetramethrin/resmethrin)
Tender Leaf Plant Insect Control (pyrethrin)
Tender Leaf Spider Mite Control (resmethrin)
Tender Leaf Systemic Granules Insect Control
 (disulfoton)
Tender Leaf Whitefly & Mealybug Control
 (resmethrin)
Termite Killer (chlorpyrifos)
Vegetable Insect Control (pyrethrin)
Vegetable Insect Killer (pyrethrin)

Herbicides

Barren Concentrate (prometon)
Brush Killer (triclopyr)
Crack & Crevice Weed Killer
 (diquat/sodium chlorate)
Ezy-Spray Weed & Feed (2,4-D/MCPP/2,4-DP)
Grass-Out (fluaziprop-butyl)
Liquid Edger (diquat)
Poison Ivy & Poison Oak Killer
 (2,4-D/MCPP/dicamba)
Pre-Emergence Weed & Grass Preventer (EPTC)
Spot Weeder (2,4-D/MCPP/dicamba)
Weed & Grass Killer (diquat)
Weed & Grass Killer Concentrate (diquat)
Weed & Grass Killer RTU (diquat)
Weed-Out (2,4-D/MCPP/dicamba)
Weed-Out Lawn Weed Control
 (2,4-D/MCPP/dicamba)

Fungicides

Bordeaux Mixture (copper sulfate)
Fungicide with Daconil 2787 (chlorothalonil)

Growth Regulators

Blossom Set for Tomatoes & Vegetables (cytokins)
Rootone Brand F Rooting Powder

Miscellaneous Chemicals

Dog & Cat Repellent (methyl nomyl ketone)
Gopher Gasser (potassium nitrate/sulfur)
Gopher Killer Pellets (zinc phosphide)
Mole Killer Pellets (zinc phosphide)
Stump Remover

DRAGON CORPORATION

P.O. Box 7311
Roanoke, Virginia 24019
(703) 362-3657
Fax (703) 362-9171

Insecticides

5% Diazinon Granules
25% Diazinon Insect Spray
25% Methoxychlor Insect Spray
50% Malathion Insect Spray
Ant Flea & Roach Killer (chlorpyrifos)
Ant Flea & Tick Insect Granule (chlorpyrifos)
Ant Killer Granules (chlorpyrifos)
Ant Roach & Spider Killer (chlorpyrifos/pyrethrin)
Blue Dragon Garden Dust (carbaryl)
Copper Dragon Tomato & Vegetable Dust
   (carbaryl/copper)
Copper Dragon Tomato & Vegetable Dust
   (carbaryl/copper)
Cygon 2E (dimethoate)
Diazinon Granules
Dipel Dust (B.t.)
Dursban 1E (chlorpyrifos)
Dursban Insect Spray (chlorpyrifos)
Fire Ant Killer (chlorpyrifos)
Flea & Tick Killer (chlorpyrifos/pyrethrin)
Flying & Crawling Insect Spray (tetramethrin)
Gentle Care House Plant Spray (pyrethrin)
Home Pest Killer (chlorpyrifos)
Hornet & Wasp Killer (propoxur)
Indoor Insect Fogger (esfenvalerate)
Lindane Borer Spray
Roach Kill (boric acid)
Rose & Flower Insect Spray (pyrethrin)
Sevin 5% Dust
Sevin 10% Dust
Sevin 50% Wettable
Snail & Slug Killer Pellets (metaldehyde)
Spider Mite & Mealybug Control (resmethrin)
Systemic House Plant Insect Control (disulfoton)
Systemic Ornamental Insect Granules (disulfoton)
Systemic Rose & Flower Care (disulfoton)
Termite Killer (chlorpyrifos)
Thiodan Insect Spray (endosulfan)
Thiodan Vegetable & Ornamental Dust (endosulfan)
Thuricide (B.t.)
Tomato & Vegetable Insect Spray (pyrethrin)
Whitefly & Mealybug Spray (resmethrin)

Herbicides

Crabgrass & Nutgrass Killer (MSMA)
Crabgrass Plus Broadleaf Weed Killer
   (MSMA/2,4-D/MCPP/dicamba)
Grass & Weed Killer (diquat)
Lawn Weed Killer (2,4-D/MCPP/dicamba)
Poison Ivy Poison Oak Killer
   (2,4-D/MCPP/dicamba)
Super Brush Killer (2,4-D/2,4-DP)
Total Vegetation Killer (prometon)

Fungicides

Bordeaux Mix (copper)
Captan Wettable
Copper Fungicide (copper-fatty & resin acids)
Daconil 2787 (chlorothalonil)
Ferbam
Fruit Tree Spray
   (captan/malathion/methoxychlor/carbaryl)
Garden Sulphur
Mancozeb Disease Control
Rose & Flower Dust
   (captan/malathion/methoxychlor/carbaryl)
Rose Flower & Insect Spray
   (captan/malathion/methoxychlor/carbaryl)
Systemic Fungicide 3336 WP (thiophanate-methyl)
Tomato Blossom End Rot Spray (calcium chloride)

Miscellaneous Chemicals

Mole Killer Pellets (zinc phosphide)
Rabbit & Dog Repellent (napthalene/dried blood)
Rootone (NAA/thiram)
Scat Cat Repellent (methyl nomyl ketone)
Tomato & Vegetable Fruit Set (cytokinin)

GREEN LIGHT COMPANY

P.O. Box 17985
San Antonio, TX 78217
(210) 494-3481
FAX (210) 494-5244

These products are NOT listed elsewhere in this book.

Insecticides

Bio-Worm Killer (B.t.)
Borer Killer (chlorpyrifos)
Diazinon 25E
Dipel Dust (B.t.)
Dormant Spray (petroleum oil)
Dursban Many Purpose Concentrate (chlorpyrifos)
Fire Ant Killer (chlorpyrifos)
Flea & Tick Granules (chlorpyrifos)
Flea & Tick Spray, Indoor RTU (chlorpyrifos)
Home Pest & Carpet Dust (bendiocarb)
Home Pest Insect Control RTU (chlorpyrifos)
Malathion 50%
Many Purpose Insect Killer (diazinon)
Red Spider Spray (dicofol)
Roach Powder (boric acid)
Rose & Flower Insect Spray RTU (pyrethrins)
Scale Away (carbaryl)
Sevin 5% Dust (carbaryl)
Sevin Granules (carbaryl)
Sevin, Liquid Flowable (carbaryl)
Striker Roach & Ant Bait (trichlorfon)
Wasp & Hornet Spray (resmethrin)
Whitefly & Mealybug Spray RTU (chlorpyrifos)

Herbicides

Amaze Granules (benefin/oryzalin)
Betasan 3.6 G (bensulide)
Com-Pleet (prometron)
Eptam 2.3% Weed & Grass Preventer (EPTC)
Eptam 5G (EPTC)
First Down Granules (benefin/trifluralin)
Liquid Edger (cacodylic acid)
MSMA Crabgrass Killer
Protrait Granules (isoxaben)
Wipe Out Broadleaf Weed Killer (2,4-D/MCPP/dicamba)
Wipe Out Broadleaf Weed Killer RTU (2,4-D/MCPP/dicamba)

Fungicides

Broad Spectrum Mancozeb Fungicide
Fung-Away Fungicide (tiadimefon)
Fung-Away Granules (tiadimefon)
Fung-Away Systemic Lawn Spray (tiadimefon)
Systemic Fungicide (thiophanate-methyl)
Wettable Dusting Sulfur

Miscellaneous

Bug & Snail Bait (carbaryl/metaldehyde)
Rat & Mouse Bait Bar (diphacinone)
Rat & Mouse Bait Place-Packs (diphacinone)
Rootone (NAA/thiram)
Tomato Bloom Spray II RTU

LAWN & GARDEN PRODUCTS INC.

P.O. Box 5317
Fresno, CA 93755
(209) 225-4770
FAX (209) 225-1319

Monterey Brand

Insecticides

Bug Buster (esfenvalerate)
Bug Buster-O (pyrethrin)
Sat-T-Side (petroleum oil)
Worm Ender (B.t.)

Herbicides

Algae Attack (copper pentahydrate)
Brush Buster (2,4-D/dichorprop)
Monterey Weed Hoe (MSMA)
Poast (sethoxydim)
Vegetable, Turf & Ornamental Weeder (DCPA)
Weed Ender (cacodylic acid)
Weed Stopper (oryzalin)
Weed Whacker (2,4-D/MCPP/2,4-DP)
Weed Whacker Jet Spray (2,4-D/MCPP/2,4-DP)

Fungicides

Bayleton 25% Fungicides (triadimefon)
Foli-Cal (calcium)
Liqui-Cop (copper-ammonia complex)
Monterey Bravo Flowable Fungicide (chlorothalonil)

Miscellaneous

Florel Brand Fruit Eliminator (ethephon)
Quintox (chlolecalciferol)
Sucker Stopper (NAA)

LILLY MILLER

14546 N. Lombard St.
Portland, Oregon 97203
(503) 289-5937
FAX (503) 289-9216

LILLY MILLER BRANDS

Insecticides

5% Diazinon Granules
Ant Flea & Spider Killer (chlorpyrifos)
Ant Flea & Tick Killer (chlorpyrifos)
Ant Killer Plus (bendiocarb)
Bug-Off Rose & Flower Spray (pyrethrin)
Cutworm, Earwig & Sowbug Bait (carbaryl)
Diazinon Insect Dust
Diazinon Insect Spray
Dursban Insect Granules (chlorpyrifos)
Dursban Insect Spray (chlorpyrifos)
Dursban Lawn Insect Granules (chlorpyrifos)
Fruit & Berry Insect Spray (diazinon)
Grasshopper Earwig & Sowbug Bait (carbaryl)
Indoor Flea Control (allethrin/chlorpyrifos)
Malathion 50%
Pestkill Rhodendron & Rose Dust (bendiocarb)
Rose & Garden Insect Fogger (pyrethrin)
Sevin 5% Dust (carbaryl)
Sevin Spray (carbaryl)
Spray Oil (petroleum oil)
Systemic Rose Care (disulfoton)
Tomato & Vegetable Dust
   (malathion/carbaryl/sulfur)
Tomato & Vegetable Insect Fogger (pyrethrin)
Whack Hornet & Wasp Killer (chlorpyrifos)

LILLY MILLER BRANDS (continued on pg 8)

MANUFACTURERS & THEIR PRODUCTS

LILLY MILLER

14546 N. Lombard St.
Portland, Oregon 97203
(503) 289-5937
FAX (503) 289-9216

LILLY MILLER BRANDS (continued)

Herbicides

Blackberry & Brush Killer (triclopyr)
Casoron Granules (dichlobenil)
Dandelion Killer (2,4-D)
Granular Noxall Vegetation Killer (sodium metaborate/sodium chlorate)
Hose & Go Moss Out (ferric sulfate)
Hose N Go Weed & Feed (2,4-D/MCPP/dicamba)
Knock-Out Weed & Grass Killer (diquat)
Knock-Out Weed & Grass Killer RTU (diquat)
Lawn Weed Killer (2,4-D/MCPP/dicamba)
Lawn Weed Killer RTU (2,4-D/MCPP/dicamba)
Moss-Kill (zinc chloride)
Moss-Kill Granules (zinc chloride)
Moss-Kill RTU (zinc chloride)
Moss-Out (ferric sulfate)
Moss-Out Granules (ferric sulfate)
Moss-Out Lawn Granules (ferrous sulfate)
Moss-Out Plus Fertilizer (ferrous sulfate)
Noxall Vegetation Killer (prometron)
Spurge & Oxalis Killer (2,4-D/MCPP/dicamba)
Super Rich Lawn Food with Moss Control (ferrous sulfate)
Super Rich Weed & Feed (2,4-D/MCPP/dicamba)
Ultragreen Crabgrass Control & Lawn Food (benefin/trifluralin)
Ultragreen Moss Control Lawn Food (ferrous sulfate)
Ultragreen Weed & Feed (2,4-D/MCPP/dicamba)
Weed & Grass Preventer (oryzalin)

Fungicides

Bulb Dust (thiram/methoxychlor)
Captan
Disease Control (chlorothalonil)
Microcop Fungicide (basic copper sulfate)
Polysul Summer & Dormant Spray (calcium poly sulphide)

Miscellaneous

Go-West Meal (metaldehyde/carbaryl)
Hose & Go Snail & Slug Killer (metaldehyde)
Rootone (NAA/thiram)
Slug & Snail Bait (metaldehyde)
Slug & Snail Line (metaldehyde)
Slug Snail & Insect Killer Bait (metaldehyde/carbaryl)
Snail & Slug Pellets (metaldehyde)
Stump Remover (potassium nitrate)

COOKE LABEL PRODUCTS

Insecticides

5% Sevin Dust (carbaryl)
50% Malathion
Ant Barrier (allethrin/chlorpyrifos)
Ant Barrier Granules (chlorpyrifos)
Diazinon Insect Granules
Diazinon Spray
Dursban Granules (chlorpyrifos)
Dursban Plus (chlorpyrifos)
Earwig Sowbug & Grasshopper Bait (carbaryl)
Garden Insect Spray (endosulfan)
Rose & Flower Dust (malathion/carbaryl/sulfur)
Rose & Flower Insect Spray (pyrethrin)
Sevin Liquid
Summer & Dormant Oil (petroleum oil)
Systemic Rose Shrub & Flower Care (disulfoton)
Tomato & Vegetable Dust (malathion/carbaryl/sulfur)

Herbicides

Broadleaf Weed & Dandelion Killer (2,4-D/MCPP/dicamba)
Brush Killer (triclopyr)
Casoron Granules (dichlobenil)
Spurge Oxalis & Dandelion Killer (2,4-D/MCPP/dicamba)
Spurge Oxalis & Dandelion Killer RTU (2,4-D/MCPP/dicamba)
Weed & Grass Preventer (oryzalin)

Fungicides

Copper Fungicide (basic copper sulfate)
Daconil Lawn & Garden Fungicide (chlorothalonil)
Fungicide (PCNB)
Kop-R-Spray (copper-ammonium complex)
Sulf-R-Spray (calcium poly sulphide)
Sulfur Dust

Miscellaneous

Gopher Mix (strychnine)
Rootone (NAA/thiram)
Slug N Snail Granules (metaldehyde/carbaryl)
Slug N Snail Spray (metaldehyde)
Snail & Slug Pellets (metaldehyde)
Stump Remover (potassium nitrate)
Tomato Plus (chlorophenoxyacetic acid)

PACE INTERNATIONAL

P.O. Box 558
Kirkland, Washington 98083
(206) 827-8711
FAX (206) 822-8261

Herbicides

Lawn Food & Weed Control (2,4-D/MCPP/2,4-DP)
Weed & Feed (2,4-D)

Molluscides

Deadline (metaldehyde)
Deadline Bullets (metaldehyde)

Repellents

Hinder Deer & Rabbit Repellent (fatty acids)

PBI GORDON

Acme Consumer Products
P.O. Box 4090
Kansas City, MO 64101
(800) 821-7925
Fax (816) 474-0462

Insecticides

2% Systemic Granules (disulfoton)
Ant Granules (chlorpyrifos)
Bagworm & Mite Spray (diazinon)
Diazinon 5% Granules
Diazinon 25% EC
Dormant Oil Spray (petroleum oil)
Dursban Granules (chlorpyrifos)
Dursban Insecticide (chlorpyrifos)
Garden Spray (pyrethrin)
Malathion 50% Spray
Proto White Grub Control (trichlorfon)
Roach Red Home Pest Killer (chlorpyrifos)
Sevin 5% Dura-Spray (carbaryl)
Sevin 50W (carbaryl)
Sevin Liquid Dura-Spray (carbaryl)
Wasp & Hornet Jet Spray (diazinon/pyrethrin)

Herbicides

Brush No More (2,4-D/MCPP/dicamba)
Crabgrass & Nutgrass Killer (MSMA)
Garden Weed Preventer Granules (DCPA)
Liquid Edger (cacodylic acid)
Super Chickweed Killer (2,4-D/MCPP/dicamba)
Trimec Lawn Weed Killer (2,4-D/MCPP/dicamba)
Trimec Plus (MSMA/2,4-D/MCPP/dicamba)
Trimec Weed & Feed 24-4-8
   (2,4-D/MCPP/dicamba)
Trimec Weed No More Spot Weeder
   (2,4-D/MCPP/dicamba)
Vegetation Killer (prometron)

Fungicides

Bordeaux Mixture (copper)
Lime Sulfur Spray (calcium polysulfide)
Liquid Fruit Tree Spray
   (captan/malathion/methoxychlor/carbaryl)
Maneb Tomato & Vegetable Fungicide
Multi Purpose Fungicide (chlorothalonil)

Miscellaneous

Mole & Gopher Killer (zinc phosphide)

MANUFACTURERS & THEIR PRODUCTS

PENNINGTON SEED INC.

P.O. Box 290
Madison, Georgia 30650
(800) 768-5400

These products are NOT listed elsewhere in this book.

Insecticides

Ant, Flea & Tick Granules (chlorpyrifos)
Ant, Flea, Tick & Grub Granules (chlorpyrifos)
Diazinon 25%
Diazinon 5% Granules
Dipel Dust (B.t.)
Dursban Extra 1% (chlorpyrifos)
Dursban Granules .5% (chlorpyrifos)
Dursban Insect Spray (chlorpyrifos)
Eliminator Fire Ant Granules
Eliminator Roach Spray
Eliminator Total Control Fire Ant (fenoxycarb)
Fire Ant Bait
Garden Dust (pyrethrin)
Garden & Ornamental Spray (pyrethrin)
Home Pest Insect Control
Hornet & Wasp
Indoor Fogger (pyrethrin)
Insect Control Plus Fertilizer
Lawn & Indoor Insect Control (chlorpyrifos)
Liquid Sevin (carbaryl)
Malathion 50%
Mole Cricket Bait (chlorpyrifos)
Roach & Ant (pyrethrin)
Rose & Flower RTU (pyrethrin)
Rose & Flower Systemic (disulfoton)
Sevin Dust 5% (carbaryl)
Sevin Dust 10% (carbaryl)
Termite Control (chlorpyrifos)
Tick, Ant & Flea Killer
Tomato & Vegetable RTU (pyrethrin)
Vegetable & Flower Spray RTU (pyrethrin)

Herbicides

Crabgrass Preventer (benefin)
Crabgrass Preventer (prodianne)
Crabgrass Preventer with Fertilizer (benefin)
Knock Out (glyphosate)
Lawn Weed Killer Granules
Liquid Edger (cacodylic acid)
Pennington0Kil (monobor chlorate)
Post-Emergence Grass Killer
St. Augustine & Centipede Weed & Feed (atrazine)
Weed & Feed w/Triamine (2,4-D/MCPP/2,4-DP)
Weed Killer Concentrate
Weed Killer RTU

Fungicides

Multi-Purpose Fungicide (chlorothalonil)
Rose & Flower Dust
  (captan/malathion/methoxychlor)
Tomato Bloom Spray (calcium)
Tomato & Vegetable Dust
  (captan/malathion/methoxychlor)

Miscellaneous

Dog & Cat Repellent
Rat & Mouse Bait Block
Rat & Mouse Killer
Rootone (IBA/thiram)
Snail & Slug Bait (metaldehyde)

RINGER CORPORATION

Safer Brand Division
9959 Valley View Rd.
Eden Prairie, MN 55344
(612) 941-4180
Fax (612) 941-5036

Insecticides

- Safer Bioneem Multi Purpose Concentrate (azadirachtin)
- Safer Flower Garden Insecticidal Soap
- Safer Fruit & Vegetable Insect Killer (insecticidal soap)
- Safer Home Patrol Insect Killer (permethrin)
- Safer Houseplant Insecticidal Soap
- Safer Indoor Insect Fogger (tetramethrin)
- Safer Insecticidal Soap Multi-Purpose
- Safer Rose & Flower Insect Killer (insecticidal soap)
- Safer Superstop Ant Roach & Insect Killer (pyrethrin/permethrin)
- Safer Tomato & Vegetable Insect Killer (insecticidal soap/pyrethrin)
- Safer Tree, Shrub & Vegetable Caterpillar Killer (B.t.)
- Safer Wasp & Hornet Killer Aerosol (tetramethrin)
- Safer Yard & Garden Insect Killer (insecticidal soap/pyrethrin)

Herbicides

- Safer Home Deck & Patio Moss & Algae Killer (fatty acids)
- Safer Lawn Moss Killer (fatty acids)
- Safer Super Superfast Brand Weed & Grass Killer (fatty acids)
- Supreme Lawn Fertilizer & Crabgrass Preventer (triflualin/benefin)
- Supreme Lawn Fertilizer & Weed Control (MCPP/MCPA/dicamba)

Fungicides

- Safer Flower Fruit & Vegetable Garden Fungicide (sulfur)

SCHULTZ COMPANY

P.O. Box 173
St. Louis, MO 63043
(314) 298-2700
(FAX) 298-2777

Insecticides

- Fruits & Vegetable Insect Spray (pyrethrin)
- House Plants & Garden Insect Spray (pyrethrin)
- Roses & Flowers Insect Spray (pyrethrin)

SECURITY BRAND

A Division of Farnam Companies
P.O. Box 34820
Phoenix, AZ 85067-4820
(602) 825-2555
FAX (602) 825-1803

These products are NOT listed elsewhere in this book.

Insecticides

Ant Killer Granules (chlorpyrifos)
Ant Killer RTU (chlorpyrifos)
Ant Killer Spray (permethrin/pyrthrin)
Cygon 2E (dimethoate)
Diazinon Spray
Diazinon Yard & Garden Granules
Dipel Dust (B.t.)
Dursban Spray (chlorpyrifos)
Flea & Tick Dip (pyrethrin)
Flea & Tick Killer for Pets (pyrethrin)
Flea & Tick Powder for Pets (carbaryl/pyrethrin)
Flea & Tick Shampoo (d-limonene)
Flea & Tick for Yards (enfenvalerate)
Home & Patio Pest Control (chlorpyrifos)
Hornet & Wasp Killer
Houseplant & Insect Control (pyrethrin)
Malathion Spray
Nature Guard for Gardens (B.t.)
Nature Guard Tomato & Vegetable Spray (pyrethrin)
Rose & Flower Insect Spray (pyrethrin)
Sevin Spray (carbaryl)
Spidermite & Mealybug Spray (pyrethrin)
Systemic Granular Insecticide (disulfoton)
Systemic Rose & Flower Care (disulfoton)
Thuricide Spray (B.t.)

Herbicides

Eze Garden Weed Preventer (trifluralin)
Killer Kare
Lawn Weed Killer (2,4-D/MCPP/dicamba)
Purge (atrazine)
Weed & Grass Killer (cacodylic acid)

Fungicides

Captan Spray
Fruit-Guard (captan/malathion)
Fungi-Guard (chlorothanil)
Lime Sulphur Solution
Nutonex Sulphur
Stop Rot (calcium)

Repellents

Detour Dog & Cat Repellent
Detour Rabbit & Deer Repellent
Detour Squirrel & Bat Repellent
Repel Animal Repellent
Repel Dog & Cat Repellent
Repel Liquid Animal Repellent
Repel Pet & Stray

Miscellaneous

Moss Master (iron sulfate)
Rootone F Rooting Compound (IBA/thiram)
Stump Remover (potassium nitrate)

Rodenticides

D-Cease (defethialone)
Just One Bit (bromadiolone)

SOUTHERN AGRICULTURAL INSECTICIDES

P.O. Box 218
Palmetto, FL 34220
(813) 722-3285
FAX (813) 723-2974

Insecticides

Cutworm & Cricket Bait (carbaryl)
Cygon 2E (dimethoate)
Diazinon 5% Granules
Diazinon 25% Insecticide
Dipel Dust (B.t.)
Dursban 1% (chlorpyrifos)
Dusrban 1E (chlorpyrifos)
Dursban Ant & Turf Granules (chlorpyrifos)
Durban Mole Cricket Bait (chlorpyrifos)
Home Pest Control (chlorpyrifos)
Home Pest Control Concentrate (chlorpyrifos)
Kelthane (dicofol)
Livestock Dust (tetrachlorvinphos)
Malathion 5% Dust
Malathion 50% EC
Malathion-Oil (malathion/petroleium oil)
Natural Pyrethrin Concentrate
Oftanol (isofenphos)
Sevin 1.75% Dust (carbaryl)
Sevin 5% Dust (carbaryl)
Sevin 10% Dust (carbaryl)
Sevin Liquid 2F (carbaryl)
Sevin 50 W (carbaryl)
Soluble Oil Spray (petroleum oil)
Thiodan .75 (endosulfan)
Thiodan 4 Dust (endosulfan)
Thuricide HPC (B.t.)

Herbicides

2,4-D Amine Weed Killer
Atrazine 4L
Balfin Granules (benefin)
Brush Killer (triclopyr)
Copper Sulfate Crystals
Lawn Weed Killer (2,4-D/MCPP/dicamba)
Surflan A.S. (oryzalin)
Weed Granules (DCPA)

Fungicides

Captan
Dithane M-45
Fruit Spray Concentrate
  (captan/malathion/methoxychlor/sulfur)
Lawn Ornamental & Vegetable Fungicide (chlorothalonil)
Liquid Copper Fungicide
  (copper salts of fatty & resin acids)
Systemic Fungicide for Turf & Ornamentals (triadimefon)
Thiomyl (thiophanate-methyl)
Tomato Dust (basic copper sulfate)
Turf Fungicide Granules (triadimefon)
Wettable or Dusting Sulfur

Mollescicides

Bait Pellets (metaldehyde/carbaryl)

SUN COMPANY INC. AGRICULTURAL PRODUCTS

Ten Penn Center • 1801 Market St.
Philadelphia, PA 19103-1699
(215) 977-3556
Fax (215) 246-8452

Insecticides

Sun Spray Ultra-Fine (petroleum oil)

MANUFACTURERS & THEIR PRODUCTS

THE SOLARIS GROUP (A DIVISION OF MONSANTO)

P.O. Box 5008
San Ramon, CA 94583-0808
(800) 225-2883

Insecticides

- Ant Stop Ant Killer Bait II (propoxur)
- Ant Stop Ant Killer Dust (chlorpyrifos)
- Ant Stop Ant Killer RTU (chlorpyrifos)
- Ant Stop Ant Killer Spray (tetramethrin/sumithrin)
- Ant Stop Logic Fire Ant Bait (fenoxycarb)
- Ant Stop Orthene Fire Ant Killer (acephate)
- Bug-B-Gon Insect Killer (diazinon)
- Diazinon Granules
- Diazinon Plus Insect Spray
- Diazinon Soil & Turf Insect Spray
- Dursban Lawn & Garden Insect Control (chlorpyrifos)
- Dursban Lawn Insect Spray (chlorpyrifos)
- Dursban Ready Spray (chlorpyrifos)
- Fire Ant Killer Granules (diazinon)
- Flea-B-Gon Flea & Tick Killer (chlorpyrifos)
- Flea-B-Gon Outdoor Flea & Tick Killer (chlorpyrifos)
- Flea-B-Gon Pet Flea & Tick Killer (pyrethrin)
- Flea-B-Gon Total Flea Killer (methoprene/permethrin)
- Home Defense Flying & Crawling Insect Killer (allethrin)
- Home Defense Hi-Power (pyrethrin/permethrin)
- Home Defense Hi-Power Roach, Ant & Spider Spray (diazinon/pyrethrin)
- Home Defense Home & Garden Insect Killer (tetramethrin/sumithrin)
- Home Defense Indoor/Outdoor Insect Killer 2 (chlorpyrifos/allethrin)
- Hornet & Wasp Killer 2 (diazinon/pyrethrin)
- Isotox (acephate)
- Malathion 50 Plus
- Orthene Systemic Insect Control (acephate)
- Ortho-Klor Insect & Termite Killer (chlorpyrifos)
- Outdoor Insect Fogger (resmethrin)
- Rose Pride Rose & Flower Insect Killer (pyrethrin)
- Rose Pride Systemic Rose & Flower Care (disulfoton)
- Sevin Carbaryl Insecticide 5 Dust (carbaryl)
- Sevin Carbaryl Insecticide 10 Dust (carbaryl)
- Sevin Garden Dust (carbaryl)
- Sevin Liquid Carbaryl Insecticide 2 (carbaryl)
- Tomato & Vegetable Insect Killer (pyrethrin)
- Volk Oil Spray (petroleum oil)
- Yard Basic Multi-Purpose Insect Killer (diazinon)

Herbicides

- Brush-B-Gon (triclopyr)
- Brush-B-Gon Poison Ivy, Oak & Brush Killer (triclopyr)
- Brush-B-Gon RTU (triclopyr)
- Casoran Granules 2 (diclobenil)
- Crabgrass Killer Formula II (CAMA)
- Crabgrass & Nutgrass Killer (CAMA)
- Fence & Yard Edger (glyphosate)
- Grass-B-Gon Grass Killer RTU (fluazifop-p-butyl)
- Green Sweep Weed & Feed (2,4-D/MCPP/2,4-DP)
- Ground Clear Triox (oxyfluorfen/imazapyr)
- Kleeraway Grass & Weed Killer RTU (glyphosate)
- Kleeraway Systemic Weed & Grass Killer (glyphosate)
- Roundup Weed & Grass Killer Concentrate (glyphosate)
- Roundup Weed & Grass Killer RTU (glyphosate)
- Roundup Weed & Grass Killer Super Concentrate (glyphosate)
- Super Edger Grass & Weed Control (glyphosate/oxyfluorfen)
- Triox (prometon)
- Weed-B-Gon (2,4-D/MCPP)
- Weed-B-Gon Jet Weeder (2,4-D/MCPP/2,4-DP)
- Weed-B-Gon Lawn Weed Killer 2 (2,4-D/MCPP/dicamba)
- Weed-B-Gon Ready Spray Lawn Weed Killer (2,4-D/MCPP)
- Weed-B-Gon for Southern Lawns (2,4-D/MCPP/dicamba)
- Yard Basics Lawn Weed & Feed (2,4-D/MCPP/2,4-DP)
- Yard Basics Lawn Weed & Feed RTU (2,4-D/MCPP/2,4-DP)
- Yard Basics Weed & Grass Killer (glyphosate/acifluorfen)
- Yard Basics Weed Killer for Lawns (2,4-D/MCPP/2,4-DP)

Fungicides

- Daconil 2787 (chlorothalonil)
- Dormant Disease Control Lime Sulfur Spray (calcium polysulfides)
- Home Orchard Spray (captan/malathion/methoxychlor)
- Rose Pride Funginex Rose & Shrub Disease Control (triforine)
- Rose Pride Orthenex Insect & Disease Control (acephate/triforine)

Mollisicides

- Bug-Geta Plus (metaldehyde/carbaryl)
- Bug-Geta Snail & Slug Pellets (metaldehyde)

TIGER PRODUCTS
CAPE FEAR CHEMICAL COMPANY

Highway 701 South
Elizabethtown, NC 28337
(910) 862-3139
FAX (910) 862-6093

These products are NOT listed elsewhere in this book.

Insecticides

Cygon 2E (dimethoate)
Diazinon 5%
Diazinon 4E
Diazinon 25% Spray
Di-Syston Systemic Granules (disulfoton)
Dormant Oil (petroleum oil)
Dursban 1E (chlorpyrifos)
Dursban 2E (chlorpyrifos)
Dursban 1% Granules (chlorpyrifos)
Fire Ant Granules (diazinon)
Flea, Tick & Chigger Killer (chlorpyrifos)
Garden Special (carbaryl/B.t.)
Home Pest Control (chlorpyrifos)
Lindane 20%
Liquid Sevin
Livestock Dust (methoxychlor/malathion)
Malathion 4% Dust
Malathion 50% Spray
New Fire Ant Granules (chlorpyrifos)
Oftanol Grub Eater (isofenphos)
Pet & Livestock Spray (pyrethrin)
Sevin 1 3/4% Dust (carbaryl)
Sevin 5% Dust (carbaryl)
Sevin 10% Dust (carbaryl)
Super Kill (endosulfan)
Sure Kill (chlorpyrifos)
Termite Killer (chlorpyrifos)
Thiodan Spray (endosulfan)
Two Way Spray (carbaryl/malathion)
Two Way Vegetable Dust (carbaryl/malathion)
Worm Killer Spray (B.t.)
Worm Whipper Dust (B.t.)

Herbicides

Brush Killer (2,4-D/2,4-DP)
Copper Sulfate
Garden Weed & Grass Preventer (trifluralin)
Kill All Liquid (sodium chlorate/sodium metaborate)
Knok-Out Grass & Weed Killer (monobor chlorate)
Lawn Weed Killer (2,4-D/dicamba)
Weed & Feed Fertilizer (2,4-D/MCPP/2,4-DP)

Fungicides

Fruit Tree Spray
    (captan/methoxychlor/malathion/carbaryl)
Rose & Flower Spray
    (captan/methoxychlor/malathion/carbaryl)
Rose Spray
    (captan/methoxychlor/malathion/carbaryl)
Sulfur 90%

Miscellaneous

Dog & Cat Repellent (methynomyl-ketone)

UNITED HORTICULTURAL SUPPLY

1407 NE Arndt Road
Aurora, OR 97002
(503) 678-9000
FAX (503) 678-9009

Growers Choice Label

Insecticides

Diazinon 5% Granules
Diazinon 25% Multi Use Insect Spray
Dursban 1% Granules (chlorpyrifos)
Dursban Lawn Insect Spray (chlorpyrifos)
Malathion 50%
Sevin 5% Dust (carbaryl)
Sevin Liquid (carbaryl)
Systemic Rose & Flower Food (disulfoton)

Herbicides

Casoran Granules (dichlorbenil)
Dandelion & Broadleaf Weed Control (2,4-D/MCPP/2,4-DP)
Kleen Up Grass & Weed Killer RTU (glyphosate/acifluorfen-sodium)
Kleen Up Spot Weed & Grass Killer (glyphosate)
Kleen Up Super Edger
Kleen Up Systemic Weed & Grass Killer (glyphosate)
Kleen Up Weed & Grass Killer (glyphosate)
Moss Killer with Lawn Food (ferrous sulfate)
Triamine Weed & Feed (2,4-D/MCPP/2,4-DP)
Weed & Feed (2,4-D/MCPP/dicamba)
Weed & Grass Killer (diquat)

UNIVERSAL INDUSTRIES SPECTRUM GROUP

P.O. Box 15842
St. Louis, Missouri 63114-0842
(800) 332-5553

These products are NOT listed elsewhere in this book.

Insecticides

6000 Lawn & Garden Insect Control (diazinon)
Bug Stop (sumithrin/tetramethrin)
Diazinon Concentrate
Fire Ant Killer (diazinon)
Home Insect Control RTU (tralomethrin)
Rose & Garden Insect Killer (diazinon)

Herbicides

Grass & Weed Killer Liquid (diquat)
Lawn Weed Killer (2,4-D/dicamba)

Miscellaneous

Snail & Slug Killer (metaldehyde)

VOLUNTARY PURCHASING GROUP INC.

P.O. Box 460
Bonham, TX 75418
(903) 583-5501

These products are NOT listed elsewhere in this book.

Ferti-Lome Labels

Insecticides

Bagworm & Tent Caterpillar Spray (malathion/methoxychlor)
Barren Flea & Roach Control (fenoxycarb)
Borer Killer (chlorpyrifos)
Bug Blaster Lawn & Garden Insecticide Killer (diazinon)
Bug Blaster RTU (diazinon)
Dipel Dust (B.t.)
Dormant Oil Spray (petroleum oil)
Kill-A-Bug Indoor/Outdoor Insect Control (chlorpyrifos)
Kill-A-Bug Plus Lawn Food (chlorpyrifos)
Lindane Spray
Mal-A-Cide Lawn & Garden Insect Control (malathion)
Ornamental & Evergreen Spray (dimethoate)
Quik-Kill Home Garden & Pet RTU (pyrethrin)
Red Spider Mite RTU (pyrethrin)
Rose Food with Systemic (disulfoton)
Scalecide (petroleum oil)
Sevin Spray (carbaryl)
Stinger Wasp & Hornet Spray (diazinon/pyrethrin)
Systemic Insect Granules (disulfoton)
Time Release Indoor/Outdoor Insect Control (pyrethrin)
Tree Borer Crystals (PDCB)
Whitefly & Mealybug Killer (resmethrin)

Herbicides

Brush Killer, Stump Killer (triclopyr)
Centipede Weed & Feed (2,4-D/MCPP/dicamba)
Crabgrass, Nutgrass & Dallisgrass Killer (MSMA)
Crabgrass Preventer & Lawn Food (trifluralin/benefin)
Lawn Food plus Moss Control (ferrous sulfate)
Over-The-Top Grass Killer (fluaziprop-butyl)
Perma-Kill Total Vegetation Killer (prometone)
Pre-Vent plus Lawn Food (isoxaben/trifluralin/benefin)
RTU Nutgrass, Poison Ivy & Vine Killer (glyphosate)
St. Augustine Weed & Feed (atrazine)
Vegetation Killer (sodium chlorate)
Weed & Feed Special (simazine)
Weed & Grass Preventer (DCPA)
Weed-Out Lawn Weed Killer (2,4-D/mecoprop/dicamba)
Weed-Out Weed Killer & Lawn Fertilizer (2,4-D/MCPP/dicamba)
Winterye & Weed Preventer for Southern Lawns (simazine)

Fungicides

Agra-San
Azalea, Camelia, Crepe Myrtle Spray (malathion/PCNB)
Blackspot & Powdery Mildew Control (copper hydroxide)
Broad Spectrum Fungicide (chlorothalonil)
Fire Blight Spray (streptomycin)
Fruit Tree Spray (malathion/captan)
Halt Systemic (thiophanate/methyl)
Lawn Food with Fungicide (PCNB)
Rose Spray (diazinon/chlorothalonil)
Triple Action Insecticide, Miticide, Fungicide (diazinon/chlorothalonil)
Wettable Dusting Sulphur

Molluscicides

Eliminate Snails & Slugs (metaldehyde)
Snail & Slug Bait (metaldehyde)

Miscellaneous

Dog-Gon Dog & Cat Repellent (methyl-nonyl-lketone)
Dog-Gon Dog & Cat Repellent Granules (methyl-nonyl-lketone)
Mole-Med Mole Repellent (castor oil)
Rabbit & Deer Repellent (ammonium soap of fatty acids)
Rooting Powder (IBA)

Natural Guard Labels

Insecticides

All Purpose Insecticidal Soap (potassium salt of fatty acids)
B.t.
Diatomaceous Earth
Dual Action Crawling Insect Control (diatomaceous earth/pyrethrin)
Insecticidal Soap RTU (potassium salt of fatty acids)
Natural Insect Spray (pyrethrin)
Neem Insecticide (azadirachtin)
Pyrethrin Concentrate
Pyrethrin Powder
Superior Oil Spray (petroleum oil)

Fungicides

Copper Fungicide (copper hydroxide)

MANUFACTURERS & THEIR PRODUCTS

VOLUNTARY PURCHASING GROUP INC.

P.O. Box 460
Bonham, TX 75418
(903) 583-5501

Hi-Yield American Brand Labels

Insecticides

American Thuricide Concentrate (B.t.)
Ant Killer Granules (diazinon)
Cygon 2E (dimethoate)
Diazinon 4E
Diazinon 5% Granules (diazinon)
Dipel Dust (B.t.)
Di-Syston Systemic Granules (disulfoton)
Dormant Spray (petroleum oil)
Dursban 1E (chlorpyrifos)
Dursban 2E (chlorpyrifos)
Dursban Spray (chlorpyrifos)
Dylox 6.2 Granules (trichlorfon)
Fire Ant Control Granules (diazinon)
Fire Ant Killer (chlorpyrifos)
Fire Ant Killer containing Logic (fenoxycarb)
Hi-Yield Indoor Fogger (pyrethrin/permethrin)
Kelthane Spray (dicofol)
Kill-A-Bug Granules (chlorpyrifos)
Kill-A-Bug RTU (chlorpyrifos)
Lindane
Malathion 55%
Malathion 5% Dust
Malathion Insect Spray
Mole Cricket Bait (chlorpyrifos)
Nem-A-Cide Nematode Control (clandosan)
Oftanol White Grub Control (isofenphos)
Roach Blaster (sumithrin)
Roach Powder (boric acid)
Sevin Brand Granules (carbaryl)
Sevin Dipel Dust (carbaryl/B.t.)
Sevin 10% Dust (carbaryl)
Sevin 5% Garden & Pest Dust (carbaryl)
Sevin 50W Insecticide (carbaryl)
Sevin Spray (carbaryl)
Termite & Soil Insect Killer (chlorpyrifos)
Thiodan Garden Dust (endosulfan)
Thiodan Spray (endosulfan)

Herbicides

American Dacthal Crabgrass & Weed Preventer (DCPA)
American Herbicide Spray with Treflan (trifluralin)
American Treflan Herbicide granules (trifluralin)
American 2,4-D Weed Killer
Atrazine Lawn Weed Killer
Betasan Granules (bensulide)
Crabgrass Control (trifluralin/benefin)
Crabgrass Preventer plus Lawn Food (siduron)
Eptam Weed & Grass Control (EPTC)
Hi-Yield Basagran Turf & Ornamental Herbicide (bentazon)
Killzall Weed & Grass Killer (glyphosate)
Lawn Weed Killer (2,4-D/MCPP/2,4-DP)
Liquid Edger (cacodylic acid)
MSMA 52% Weed Killer
Poast Herbicide (sethoxydrin)
Spot Weed Killer RTU (2,4-D/MCPP/2,4-DP)
Total Vegetation Control (cacodylic acid)
Weed & Feed (2,4-D/MCPP/dicamba)

Fungicides

American Copper Fungicide (copper hydroxide)
Bordeaux Mix Fungicide (copper sulfate/lime)
Captan 50W Fungicide
Consan 20 (N-alkyl ammonium chloride)
Daconil Lawn, Vegetable & Flower Fungicide (chlorothalonil)
Dusting Wettable Sulphur
Fruit Tree Spray (malathion/methoxychlor/captan)
Lawn Fungicide Granules (triadimefon)
Lime Sulphur Spray
Liquid Sulphur
Maneb Lawn & Garden Fungicide
Rose, Floral & Vegetable Dust (carbaryl/sulfur)
Terrachlor Granular Fungicide (PCNB)

Molluscicides

Snail & Slug Pellets (metaldehyde)

Miscellaneous

Mole & Gopher Bait (zinc phosphide)
Root Killer (copper sulfate)
Stump Remover (potassium nitrate)

A Guide To Lawn, Garden & Home Pest Control Products

Section II	Products & What They Control	
	Insecticides	20
	Herbicides	27
	Fungicides	32
	Miscellaneous Chemicals	36

INSECTICIDES

ACEPHATE
Orthene Fire Ant Killer (75%) – Solaris
Orthene Systemic Insect Control (9.4%) – Solaris

Fire ants, ants, greenbug, armyworm, leaf hoppers, sod web worms, mole crickets, bagworms, loopers, black vine weevil, budworms, case bears, sphinix moth, leaftier, thrips, leaf beetles, canterworms, webworms, mapleworm, gypsy moth, hornworm, lacebugs, leafhoppers, leafminers, leafrollers, maple shoot moth, mealy bug, Nantucket pine tip moth, oakworm, root weevil, tentmaker, psyllids, rose midge, sawflies, scales, spittlebug, sunflower moth, tent caterpillar, leaf beetle, mites (suppression), Tussock moth, whiteflies, yellow necked caterpillar.

ACEPHATE/HEXAKIS
Isotox Insect Killer Formula IV (8.5%) – Solaris

Aphids, grasshoper, Japanese beetle, mites, thrips, mealybugs, whiteflies, scales, plantbugs, armyworms, leafminers, black vine weevil, budworms, looprs, sphinix moth, leaf beetles, cankerworms, webworms, gypsy moth, lacebugs, leafrollers, leafhoppers, maple shoot moth, Nantucket pine tip moth, oakworms, root weevil, tent maker, psylids, sawflies, spittlebugs, stinkbugs, tent caterpillars, willow leaf beetle.

AZADIRACHTIN
Bioneem (.09%) – Ringer
Japanese Beetle Repellent (.09%) – Ringer

Aphids, bagworms, beetles, budworms, caterpillare, leafhoppers, leafminers, thrips, whiteflies, elm leaf beetle, grasshopper, gypsy moth, hornworms, pine sawflies, psyllids, tent caterpillars, webworms, weevils, armyworms, fruit flies, loopers, mealybugs, Mexican bean beetle, bilbugs, chinchbugs, crane flies, hyperodes weevils, mole crickets, sod webworms, white grubs, box elder bugs, lacebugs, Japanese beetle, bean leaf beetles, bollworms, codling moth, Colorado potato beetle, corn earworms, cucumber beetles, cutworms, diamondback moth, European corn borer, grape leaf skeletonizer, imported cabbageworm, melon worms, pickleworms, pinworms, squash bugs, tomato fruitworm, twig girdlers, vine borers, weevils.

BACILLUS THURINGIENSIS
Bacullus Thuringiensis (B.t.) (.8%) – Bonide
Bio-Worm Killer (.8%) – Green Light
Caterpillar Killer (1.76%) – Ringer
Dipel Dust (.04%) – Southern Ag
Dipel Dust (.048%) – Dragon
Dipel Dust (.064%) – Green Light
Dipel Insect Control (1%) – Dexol
Thuricide (.8%) – Dragon
Thuricide HPC (.9%) – Southern Ag

Canterworms, webworms, tent caterpillar, oak moth, red humped caterpillar, gypsy moth, grape leaf folder, Orange dog, Leafrollers, rindworms, cabbage looper, imported cabbageworm, diamond back moth, tomato fruitworm, spanworms, cabbageworms, hornworm, budworms, cankerworm.

BACILLUS THURINGIENSIS STRAIN EG-2371
Worm-Ender (10%) – Lawn & Garden Products

Armyworms, azalea moth, diamondback moth, hornworms, loopers, oleander moth, leaf rollers, loopers, budworms, bagworm, browntail moth, oakworm, tussock moth, spanworms, webworms, mapleworm, gypsy moth, pine butterfly, caterpillars, canker worm, tortrex moth, tent caterpillar, navel orangeowrm, oriental fruit moth, peach twig borer, cutworm, smorbia, orangedog, sphinx moth, grape berry moth, grape leaf skeletonizer, fruitworms, banana skipper, hornworms, celery leaftier, corn earworm, European corn borer, cabbagworm, melonworm, pickleworm, rindworm, tomato fruitworm, tomato pinworm.

B.T.VAR SAN DIEGO
Colorado Potato Beetle Beater (5.6%) – Bonide

Elm leaf beetle, Colorado potato beetle.

BENDIOCARB
Flea Beater (1%) – Bonide
Intercept H&G Grub Control Granules (2.5%) – Agr Evo
Ant Killer Plus (1%) – Lilly Miller
Pest Kil (1%) – Lilly Miller
Home Pest & Carpet Dust (1%) – Green Light
Predator Carpenter Ant Killer Dust (1%) – Dexol
Predator Termite Killer Dust (1%) – Dexol

White grubs, mole crickets, chinchbugs, leafhoppers, webworms, ants, ticks, centipedes, crickets, earwigs, firebrats, fleas, ground beetles, millipedes, pill bugs, scorpions, silver fish, sowbugs, spiders, ants, brown dog ticks, cockroaches, ground beetles, centipedes, wasps, bees, ticks, bagworm, caterpillars, citrus black fly, aphids, webworms, elm leaf beetle, whitefly, Japanese beetle, lacebugs, mealybugs, root weevil, spittlebug, scales, cankerworm, willowleaf beetle, thrips, blackvine weevil.

BORIC ACID
Predator Roach Powder (99%) - Dexol
Roach Powder (99%) - Green Light

Roaches, carpenter ants, ants, palmetto bugs, silver fish, waterbugs.

CARBARYL
5% Sevin Garden Dust - Rigo
Blue Dragon Garden Dust (2%) - Dragon

INSECTICIDES

Cutworm & Cricket Bait (5%) - Southern Ag
Cutworm, Earwig & Sowbug Bait (5%) - Lilly Miller
Earwig, Sowbug & Grasshopper Bait (5%) - Lilly Miller
Grasshopper, Earwig & Sowbug Bait (5%) - Lilly Miller
Liquid Flowable Sevin (21.3%) - Green Light
Liquid Sevin (22.5%) - Bonide
Liquid Sevin (22.5%) - Rigo
Scale Away (21.3%) - Green Light
Sevin 1.75% Dust - Southern Ag
Sevin 5 Dust (5%) - Agr Evo
Sevin 5 Dust - UHS
Sevin 5% Dura-Dust - PBI Gordon
Sevin 5% Dust - Bonide
Sevin 5% Dust - Dexol
Sevin 5% Dust - Dragon
Sevin 5% Dust - Green Light
Sevin 5% Dust - Lilly Miller
Sevin 5% Dust - Southern Ag
Sevin 10 Dust - Agr Evo
Sevin 10 Dust - Solaris
Sevin 10% - Bonide
Sevin 10% Dust - Dragon
Sevin 10% Dust - Southern Ag
Sevin 50 WP - Dragon
Sevin 50% WP - PBI Gordon
Sevin 50% WP - Rigo
Sevin 50W - Southern Ag
Sevin Garden Dust (5%) - Solaris
Sevin Granules 6.3% - Green Light
Sevin Insecticide 5 Dust - Solaris
Sevin Liquid (21.3%) - Dragon
Sevin Liquid (22.5%) - Lilly Miller
Sevin Liquid 2F (21.3%) - Southern Ag
Sevin Liquid Carbaryl Insecticide 2 (21.3%) - Solaris
Sevin Liquid Dura Spray (23.4%) - PBI Gordon
Sevin Liquid Insect Killer II (21.3%) - Dexol
Sevin Liquid Insecticide (11.7%) - UHS
Sevin Spray (21.3%) - Lilly Miller

Flea beetles, corn earworm, lygusm stinkbugs, spittlebugs, armyworms, imported cabbageworm, diamondback moth, harlequin bugs, pickleworms, cucumber beetle, tomato fruitworm, tomato hornworm, Colorado potato beetle, stinkbugs, leafroller, strawberry weevin, fleas, ticks, lice, fowl mite thrips, cutowrm, chinchbug, corn rootworm, European corn borer, tomato pinworm, tarnished plant bug, melonworm, leaf fodders, European fruit lecaniums, Western grape leaf skeletonizer, grape berry moth, leafrollers, rust mites, mealybugs, ants, apple maggots, scales, green fruitworm, black cherry aphids, cherry fruitworm, earwigs, pear scales, pear leaf blisher mite, pear rust mite, plum curalios, wooly apple aphids, codling moth, peachtree borer, peach twig borers, Tussock moth, mealy plum aphids, blueberry maggot, cranberry fruit worm, cranberry fireworm, California orang dogs, citrus root weevils, orange tortix, pecan leaf phylloxeras, filbert aphid, pecan ant casebearer, pecan weevil, walnut caterpillar, bagworms, boxelder bugs, blisher beetle, elm leaf beetle, willow leaf beetle, tent caterpillar, June beetle, gypsy moth, birch leafminers, ants, billbugs, centipedes, chiggers, fleas, grasshoppers, millipedes, mosquitoes, sod webworms, sowbugs, springtails, cankerworms, Oriental fruit moth, cat facing bugs, bagworms.

CHLORPYRIFOS

Ant Barrier (.5%) - Lilly Miller
Ant Flea & Tick Insect Killer (1%) - Dragon
Ant Flea & Tick Killer (1.7%) - Lilly Miller
Ant Granules (.5%) - PBI Gordon
Ant Killer Dust II (1%) - Dexol
Ant Killer Granules (.5%) - Dexol
Ant Killer Granules (.5%) - Dragon
Ant Roach & Spider Killer II (.5%) - Dexol
Ant Stop Ant Killer (.5%) - Solaris
Ant Stop Ant Killer Dust (1%) - Solaris
Ant, Flea & Spider Killer (.5%) - Lilly Miller
Borer Killer II (12.9%) - Green Light
Carpenter Ant Control (.5%) - Bonide
Carpenter Ant Killer (12.6%) - Dexol
Carpenter Ant Killer (12.6%) - Rigo
Dursban 1% - Southern Ag
Dursban 1% Granules - UHS
Dursban 1E (12.6%) - Southern Ag
Dursban 6.6% - Rigo
Dursban Ant & Turf Granules (.5%) - Southern Ag
Dursban Granular Insecticide (.5%) - PBI Gordon
Dursban Granules (.5%) - Lilly Miller
Dursban Granules (1%) - Dexol
Dursban Insect Spray (6.79%) - Lilly Miller
Dursban Insecticide (6.62%) - PBI Gordon
Dursban Insecticide Concentrate (12.6%) - Agr Evo
Dursban Lawn & Perimeter Granules (1%) - Agr Evo
Dursban Lawn Insect Control (1%) - Rigo
Dursban Lawn Insect Granules (1%) - Lilly Miller
Dursban Lawn Insect Killer (6.7%) - Dexol
Dursban Lawn Insect Spray (5.3%) - Solaris
Dursban Lawn Insect Spray (6.7%) - UHS
Dursban Mose Cricket Bait (.5%) - Southern Ag
Dursban Plus Lawn Insecticide (6.79%) - Lilly Miller
Dursban Ready Spray Outdoor Flea & Tick Killer (5.3%) - Solaris
Fire Ant Granules (.5%) - Agr Evo
Fire Ant Killer (.5%) - Dragon
Fire Ant Killer (6.27%) - Green Light
Fire Ant Killer Granules III (.5%) - Dexol
Flea & Tick Granules (.5%) - Green Light

INSECTICIDES

Flea & Tick Spray for Yard & Kennel (6.7%) - Dexol
Flea, Tick & Mange Dip (4.85%) - Agr Evo
Flea-B-Gon Flea & Tick Killer (.5%) - Solaris
Flea-B-Gon Outdoor Flea & Tick Killer (5.3%) - Solaris
Home Pest Control (.5%) - Bonide
Home Pest Control (.5%) - Southern Ag
Home Pest Control (6.7%) - Dexol
Home Pest Control Concentrate (10%) - Bonide
Home Pest Control Concentrate (6%) - Southern Ag
Home Pest Insect Control (.2%) - Green Light
House Plant Spray II (.06%) - Green Light
Indoor Flea & Tick Spray (.5%) - Green Light
Many Purpose Dursban Concentrate (6.6%) - Green Light
Ortho-Klor (12.6%) - Solaris
Primacide Fire Ant Dust (1%) - Agr Evo
Roach-Rid Home Pest Control (.5%) - PBI Gordon
Termi-Chlor (12.6%) - Agr Evo
Termite & Carpenter Ant Control (12.6%) - Bonide
Termite Killer (12.6%) - Dexol
Termite Killer (12.6%) - Dragon
Termite Killer (12.6%) - Rigo
Whitefly & Mealybug Spray (.06%) - Green Light

Fleas, Brown dog ticks, carpet beetles, flies, mosquitos, ants, crickets, earwigs, armyworm, chiggers, chinch bugs, clover mites, cutworm, grasshopper, sodweb worms, turfgrass weevil, aphids, bagworm, tent caterpillar, oakworm, spittlebug, mites, whiteflies, caterpillars, leaf hoppers, mealybugs, minosa webworm, red humped caterpillar, thrips, ash borer, lilac borer, scales, oak lecanium, white grubs, sowbugs, turfgrass weevil, termites, powder post beetles, house beetles, furniture beetles, death watch beetle, carpenter bees, onion maggots, hyperodes weevils, pillbugs, cutworms, root maggots, ants, fire ants, roundheaded house borers, bees, fleas, ticks, mange mites, European crane fly, elm bank beetle, sweet potato beetle, wireworms, root maggots, onion maggot, Japanese beetle, hyperodes beetle, grasshoppers, whiteflies, mealybug, scales, borers, peach tree borer, ash borer, lilac borer.

CHLORPYRIFOS/ALLETHRIN

Ant Barrier (.054%) - Lilly Miller
Home Defense Indoor Outdoor Insect Killer2 (.81%) - Solaris
Indoor Flea Killer (.69%) - Lilly Miller
Whack Hornet & Wasp Killer (.3%) - Lilly Miller

Ants, fleas, spiders, roaches, ticks, spiders, wasps, hornets, yellow jacket, tent caterpillars, gypsy moth, earwigs, silverfish, mosquitoes, crickets, clover mites, firebrats, scorpions, confused flour beetle, saw toothed grain beetle, Indian meal moth, rice weevil, flour moths, carpet beetles.

CHLORPYRIFOS/RESMETHRIN

Ultra S.S.C. (14.5%) - Agr Evo

Cockroaches, crickets, silverfish, carpet beetles, clover mites, centipedes, milepedes, firebrats, earwigs, sowbugs, ants, spiders, fleas, ticks, clover mites, pill bugs.

CYPERMETHRIN

Vikor Dual Action Home & Patio Spray (.55% a.i.) - Agr Evo
Vikor Insect Concentrate (.26%) - Agr Evo

Cockroaches, waterbugs, crickets, silverfish, firebrats, centipedes, milepedes, earwigs, sowbugs, pillbugs, spiders, ants, flies, mosquitoes, gnats, flying moths, wasps, hornets, yellow jackets, bees, spiders, fruit flies.

DIAZINON

5% Diazinon Dust - Rigo
5% Diazinon Granules - Dragon
5% Diazinon Granules - Lilly Miller
5% Diazinon Granules - Rigo
5% Diazinon Granules - Southern Ag
25% Diazinon - Rigo
25% Diazinon Insect Spray - Dragon
Ant Dust (1%) - Bonide
Ant Killer Powder (5%) - Rigo
Ant Roack & Spider Control (.5%) - Dexol
Bagworm & Mite Spray (25%) - PBI Gordon
Bug-B-Gon (.075%) - Solaris
Diazinon 2% Granules - Dexol
Diazinon 4E - Agr Evo
Diazinon 5% Granules - Agr Evo
Diazinon 5% Granules - Dexol
Diazinon 5G - PBI Gordon
Diazinon 12 1/2% E - Bonide
Diazinon 25 - Green Light
Diazinon 25% - Agr Evo
Diazinon 25% - PBI Gordon
Diazinon 25% - UHS
Diazinon 25% Insect Spray - Dexol
Diazinon Dust (4%) - Lilly Miller
Diazinon Granules (2%) - Dragon
Diazinon Granules (2%) - Solaris
Diazinon Insect Control (25%) - Bonide
Diazinon Insect Granules (5%) - Lilly Miller
Diazinon Insecticide (25%) - Southern Ag
Diazinon Plus Insect Spray (25%) - Solaris
Diazinon Soil & Turf Insect Control (5%) - Solaris
Diazinon Spray (16.75%) - Lilly Miller
Fire Ant Killer Granules (5%) - Rigo
Fire Ant Killer Granules (5%) - Solaris
Fruit & Berry Insect Spray (16.75%) - Lilly Miller
Garden Soil Insecticide 5% G - Bonide
Many Purpose Insect Killer (5%) - Green Light
Mole Cricket Killer (5%) - Bonide
Multi-Purpose Insect Killer I (.075%) - Solaris

INSECTICIDES

Ants, armyworms, bermudagrass mites, billbugs, ticks, chiggers, chinch bugs, clover mites, crickets, cutworms, fleas, digger wasps, earwigs, lawn moths, leaf hoppers, millipedes, sod webworms, sowbugs, spring tails, white grubs, Japanese beetle, European chafer, Southern chafer, mole crickets, onion maggots, root maggots, wireworms, aphids, bagworms, European pine short moth, holly budmoth, leafminers, mealybugs, mites, leaftiers, root weevils, scale, whiteflies, skeletonizer, leaf rollers, pear slugs, tent caterpillars, flea beetles, fruit flies, fruit worms, grape berry moth, grape leaf folders, twig borers, cucumber beetle, grasshoppers, Mexican bean beetle, diamondback moth, imported cabbageworm, thrips, Colorado potato beetle, vinegar flies, white grubs, wireworms, Hyperades weevil, box elder bugs, cockroaches, spiders, fire ants, pineshoot moth, bud moths.

DIAZINON/PYRETHRIN

Home Defense Hi Power Roach Ant & Spider Killer (.55%) - Solaris
Hornet & Wasp Killer 2 (.55%) - Solaris
Wasp & Hornet Jet Spray (diazinon/pyrethrin) - PBI Gordon

Wasps, hornets, digger wasps, clover mites, roaches, ants, spiders, crickets, silverfish, fleas, ticks, carpet beetle, earwigs, firebrats, stored product insects.

DICOFOL

Kelthane (42%) - Southern Ag
Kelthane Miticide (35%) - Bonide
Red Spider Spray (42%) - Green Light

Mites.

DIMETHOATE

Cygon 2E (23%) - Southern Ag
Cygon 2E (23.4%) - Rigo
Cygon 2E Systemic Insecticide (23.4%) - Dragon
Cygon Systemic Insecticide (23.4%) - Bonide
Systemic Shrub & Flower Insecticide (23.4%) - Dexol

Lace bugs, leafminers, mites, scales, whiteflies, aphids, leafhoppers, houseflies, thrips, mealybugs, bagworms, midges, European pine shoot moth, Nantucket pine tip moth, Zimmerman pine moth, pepper maggots, lygusbugs.

DISULFOTON

2% Systemic Granules - PBI Gordon
Feed & Shield Systemic Rose Care (1%) - Lilly Miller
Rose & Flower Care (1%) - Solaris
Rose Guard (1%) - Rigo
Systemic Granules (1%) - Bonide
Systemic Granules (1%) - Dexol
Systemic Granules (2%) - Bonide
Systemic House Plant Insect Control (1%) - Dragon
Systemic Rose & Flower Care (.5%) - Bonide
Systemic Rose & Flower Care (1%) - Dragon
Systemic Rose & Flower Food (.5%) - UHS
Systemic Rose Care (1%) - Dexol
Systemic Rose Shrub & Flower Care (1%) - Lilly Miller

Aphids, lacebugs, leafhoppers, leafminers, spider mites, thrips, whiteflies, fleabeetles, webworm, pinetip moth, scales, elm leaf beetles, mealybug, potato psyllid.

ENDOSULFAN

Garden Insect Spray (9.15%) - Lilly Miller
Thiodan .75 Insect Spray - Southern Ag
Thiodan 4 Dust - Southern Ag
Thiodan Insect Spray (9.9%) - Dragon
Thiodan Vegetable & Ornamental Dust (3%) - Dragon

Aphids, bean beetle, Colorado potato beetle, bean leaf skeletonizer, cucumber beetle, stink bug, cutworm, cabbage looper, cabbage worm, diamondback moth, flea beetle, harlequin bug, armyworm, crown mite, melonworm, pickleworm, rindworm, squash beetle, squash bug, squash vine borer, leaf roller, whitefly, hornworm, pepper maggot, leaf footed bug, tuber worm, psyllid, European corn borer, chinch bug, three lined potato beetle, blister beetle, hornworm, tomato russet mite, fruitworm, apple rust mite, blueberry bud mite, plum rust mite, peach twig borer, peach silver mite, peach tree borer, cyclomen mite, tarnished plant bug, rose chafer, Zimmerman pine moth, dogwood borer, box elder bug, taxus bud mite.

ESFENVALERATE

Bug Buster (.425%) - Lawn & Garden Products

Caterpillars, loopers, armyworms, artichoke plume moth, buckthorn aphid, carrot weevil, chinch bug, Colorado potato beetle, corn earworm, corn rootworm, curculio, cucumber beetle, cutworm, diamondback moth, European corn borer, flea beetle, grasshopper, cloverworm, imported cabbageworm, leafhopper, lygus, Mexican bean beetle, pea aphid, pea weevil, pepper weevil, pickleworm, psyllid, potato tuberworm, rindworm, sap beetle, corn borer, squash bug, squash vine borer, stalk borer, stink bug, tarnished plant bug, tobacco hornworm, tomato fruitworm, tomato pinworm, leafminer, cutworm, whiteflies, ants, cockroaches, crickets, palmetto bugs, sowbugs, spiders, ticks.

FATTY ACIDS–POTASSIUM SALTS

Flower Garden Insecticidal Soap (49%) - Ringer
Fruit & Vegetable Insect Killer (2%) - Ringer

INSECTICIDES

Houseplant Insecticidal Soap (2% and 49%) - Ringer
Insecticidal Soap Concentrate (.5%) - Bonide
Insecticidal Soap for Fruit & Vegetable (2%) - Bonide
Insecticidal Soap Multi Purpose Insect Killer (2% and 49%) - Ringer
Rose & Flower Insect Killer (1.5%) - Ringer

Aphids, mealy bugs, whiteflies, scales, mites, earwigs, grasshopper, leafhopper, plant bugs, psyllids, sawflies, squash bugs, harlequin bugs, adegids, lacebugs, thrips, springtail.

FATTY ACIDS/PYRETHIUM

Tomato & Vegetable Insect Killer - Ringer
Yard & Garden Insect Killer - Ringer

Aphids, asparagus beetles, bean beetles, cabbage loopers, caterpillars, Colorado potato beetle, cucumber beetle, diamondback moth, flea beetles, cabbageworms, leaf hoppers, plant bugs, hornworms, whiteflies, bagworms, gypsy moth, Japanese beetle.

FENOXYCARB

Logic Fireant Bait (1%) - Solaris

Fireants.

ISOFENPHOS

Oftanol (1.5%) – Southern Ag

Mole crickets, fire ants, grubs, billbug, hyperodes weevil, sod webworm, chinch bug.

LINDANE

Borer Miner Killer 5 (5%) - Bonide
Lindane Borer Spray (20%) - Dragon

Aphids, thrips, leafminers, cucumber beetle, lacebugs, flat-head borers, roundhead borers, peach tree borers, dogwood borers, rhododendron borers, lilac borer, pine bark beetle, turpentine beetle, ticks, fleas, chiggers, leafhoppers.

MALATHION

5% Malathion Dust - Southern Ag
50% Malathion - Agr Evo
50% Malathion - Lilly Miller
50% Malathion - Rigo
50% Malathion Insect Spray - Dragon
Malathion 50 Plus - Solaris
Malathion 50% - Southern Ag
Malathion 50% - UHS
Malathion 50% EC - Bonide
Malathion 50% Spray - PBI Gordon
Malathion Insect Control (50%) - Dexol

Mites, aphids, scales, Mexican bean beetle, leafhoppers, imported cabbageworm, flies, mosquitoes, whiteflies, thrips, mealy bugs, Japanese beetle, plantbugs, leafhoppers, mealybugs, tarnished plant bug, bagworms, pear psyllids, leaf miners, pine shoot moth, scales, codling moth, plum cuculio, leaf rollers, ornamental fruit moth, pecan shylloxera, pecan nut casebearers, lacebugs, tent caterpillar, thrips, cucumber beetle, loopers, budmoth, orange tortrix, fruit lecanium, chafers, mealybug, cucumber beetle, pickleworm, vine borer, crickets, lygus, spittlebug, thrips, asparagus beetle, blister beetle, bean leaf beetle, imported cabbageworm, diamondback moth, webworm, fleabeetles, onion maggot, grasshopper, chinchbug, pepper maggot, fruitworm, armyworm, Drosophila.

MALATHION/LINDANE/CARBARYL/DICOFOL

Mite & Insect Spray (20.5%) - Bonide

Bagworm, mites, Japanese beetle, aphid, lace bugs, leafminers, thrips, armyworms, fleabeetles, plantbugs, whiteflies, spittlebugs, caterpillars, rose chafers, taxus weevils, psyllids, grasshoppers, tarnished plantbug, pine shoot moth, scales, flies, mosquito, ants, spiders, chiggers, ticks, box elder bugs, elm beetles, wire worms, strawberry root weevil, Fuller rose beetle, cucumber beetle.

MALATHION/METOXYCHLOR

Mosquito Control Concentrate - Rigo

Mosquitoes.

MALATHION/PETROLEUM OIL

Malathion Oil (80%) – Southern Ag

Scales, whiteflies, mealybug, spider mites.

METHOPRENE

Alter Flea Control EC (1%) - Agr Evo

Fleas.

METHOXYCHLOR

25% Methoxychlor Insect Spray - Dragon
Methoxychlor 25% E - Bonide

Cankerworms, gypsy moth, Tussock moth, hemlock looper, imported willow leaf beetle, rose chafer, leafhoppers, spittlebugs, Japanese beetle, lacebug, fleabeetle, apple maggots, codling moth, plum curculios, cherry fruitworm, cherry fruit flies, strawberry weevils, fruitworms, grape leaf skeletonizer, leaf beetles, corn earworm, armyworms, Mexican bean beetle, imported cabbageworm, pea weevils, blister beetles, Colorado potato beetle, tomato fruitworm, mosquitos, gnats.

INSECTICIDES

PERMETHRIN

Home Patrol Insect Killer (.25%) - Ringer
Inter Cept Insect Control, Vegetable, Lawn & Garden Spray (13.3%) - Agr Evo
Mite Beater (.5%) - Bonide
P.C. Pest Control Concentrate (13.3%) - Bonide
Permethrin Turf & Ornamental (13.3%) - Bonide

Codling moth, pear psylla, cluster flies, tent caterpillars, ants, cockroaches, spiders, fleas, silverfish, earwigs, waterbugs, palmetto bugs, crickets, flies, wasps, hornets, yellow jackets, bees, scorpions, millipedes, centipedes, sowbugs, pillbugs, silverfish, firebrats, clover mites, cheese mites, granary weevil, rice weevil, confused flour beetle, rust red flour beetle, drugstore beetle, meal worms, grain mites, cadellas, fleas, ticks, carpet beetles, bedbugs, clothes moth, lice, whiteflies, mites, aphids, thrips, lacebugs, leafminers, clover mites, cabbageworms, fungus gnats, cankerworms, bagworms, leafrollers, inchworms, pine moths, needle sides, budworms, oakworm, webworms, Tussock moth, elm leaf beetle, elm spanworm, armyworm, Japanese Beetle, grasshoppers, cutworm, chickbugs, mole crickets, wrebworm, grasshoppers, pine beetle, oakworm, pine breetles, citrus black flires, elm spanworm, elm leaf beetle, fungus gnats, cabbageworm, corn earworm, pine moth, pine needle miners, inchworms, leaf rollers, mealybugs, cankerworm, asparagus beetle, lygus, tarnished plang bug, looper, diamondback moth, Colorado potato beetle, tobacco budworm, flea beetle, tuberworm, European corn borer, psyllids, pepperweevil, cloverworm, hornworm, pinworm, fruitworm, navel orangeworm, peach twig borer, plum curculio, oriental fruit moth.

PERMETHRIN/METHOPRENE

Alter Flea Control (.257%) - Agr Evo
Flea-B-Gon Total Indoor Fogger (.66%) - Solaris
Flea-B-Gon Total RTU (.257%) - Solaris

Fleas, ticks, roaches, silverfish, earwigs, ants.

PERMETHRIN/PYRETHRIN

Home Defense Hi-Power (.885%) - Solaris
Superstop Ant Roach & Insect Killer - Ringer

Cockroaches, waterbugs, palmetto bugs, crickets, scorpions, spiders, milipedes, centipedes, silverfish, sowbugs, pillbugs, fleas, ticks, lice, flies, fruit flies, carpet beetles, mosquitos, gnats, wasps, hornets, flying moth, clothes moth, bedbugs, yellow jackets, mole crickets.

PETROLEUM OILS

Dormant & Summer Oil Spray II (98%) - Dexol
Dormant Oil Spray (97%) - PBI Gordon
Dormant Spray & Summer Oil Spray (97%) - Green Light
Dormant Spray (98.8%) - Rigo
Horticultural Spray Oil (98%) - Bonide
Horticultural Spray Oil (98%) - Dragon
Mosquito Larvicide (98%) - Bonide
Saf-T-Side (80%) - Lawn & Garden Products
Soluble Oil Spray (98%) - Southern Ag
Spray Oil (99%) - Lilly Miller
Summer & Dormant Oil (99%) - Lilly Miller
Sun Spray Ultra Fine Spray Oil (98.8%) - Sun Oil
Volck Oil Spray (97%) - Solaris

Mites, pear psylla, scales, aphids, mealybug, plant bugs, white flies, pysllids, leafminers, sawflies, thrips, fungus gnats, leafhoppers, certain caterpillars, loopers, apple redbug, leafrollers, greasy spot, sooty mold, black fly, mosquito larvae.

PROPOXUR

Ant Stop Ant Killer Bait 2 (.25%) - Solaris

Ants.

PHOSMET

Imidan (12.5%) - Bonide

Cherry fruitfly, elm spanworm, grape berry moth, leaf hoppers, leaf fodders, gypsy moth, aphids, apple maggot, codling moth, leaf rollers, pear psylla, tarnished plant bug, oriental fruit moth, peach twig borer, mites, plum curculio, elm spanworm, cankerworm.

PHENOTHRIN/ALLETHRIN

Home Defense Flying & Crawling Insect Killer (.5%) - Solaris

Ants, roaches, flies, spiders, earwigs, sowbugs, centipedes, flies, mosquitoes, wasps, gnats.

PYRETHRIN

Bug Buster-O (1.4%) - Lawn & Garden Products
Bug Off Rose & Flower Spray (.02%) - Lilly Miller
Flea Free Carpet Treatment (.15%) - Dexol
Flea-B-Gon Pet Flea & Tick Killer (.1%) - Solaris
Fruits & Vegetable Insect Spray (.02%) - Schultz
Garden Spray (.02%) - PBI Gordon
Gentle Care House Plant Spray (.02%) - Dragon
House Plant & Garden Insect Spray (.02%) - Schultz
Household Insect Control (.1%) - Dexol
Japanese Beetle Killer (.02%) - Bonide
Natural Pyrethrin Concentrate (.96%) - Southern Ag
Organic Greenhouse, House & Vegetable Spray Concentrate (1%) - Bonide
Plant Insect Control (.02%) - Dexol
Rose & Floral Insect Control (.02%) - Dexol
Rose & Floral Insect Killer (.02%) - Dexol
Rose & Flower Insect Killer (.02%) - Solaris
Rose & Flower Insect Killer II (.02%) - Rigo

PRODUCTS & WHAT THEY CONTROL

INSECTICIDES

Rose & Flower Insect Spray (.02%) - Green Light
Rose & Flower Insect Spray (.02%) - Schultz
Rose & Flower Spray (.02%) - Lilly Miller
Rose & Garden Insect Fogger (.05%) - Lilly Miller
Rose Flower & Ornamental Insect Spray RTU (.02%) - Bonide
Tomato & Vegetable Fogger (.05%) - Lilly Miller
Tomato & Vegetable Insect Killer (.01%) - Solaris
Tomato & Vegetable Insect Spray (.02%) - Dragon
Tomato & Vegetable RTU (.02%) - Bonide
Tomato Pepper Vegetable Spray RTU (.02%) - Bonide
Vegetable Insect Control (.02%) - Dexol
Vegetable Insect Killer (.02%) - Dexol

Asparagus beetle, aphids, leafhoppers, whiteflies, Mexican bean beetle, cucumber beetles, cabbage looper, cabbageworms, stinkbugs, leaftiers, fireworms, beetles, blister beetles, Colorado potato beetle, webworms, thrips, fleabeetles, flies, fruit flies, mosquitos, gnats, ants, armyworms, asparagus beetle, loopers, caterpillar, cockroaches, corn earworms, crickets, diamondback moth, leaf rollers, leafhoppers, thrips, gypsy moth, Harlequin bug, lice, Mexican bean beetle, psyllids, skippers, stikbugs, webworms, whiteflies, mealybugs, mites, gnats, mosquitoes, moths, fer scales, leaftiers, fireworms, Japanese beetle.

RESMETHRIN

Aphid Mite & Whitefly Killer II (.057%) - Dexol
Outdoor Insect Fogger (1.284%) - Solaris
Spider Mite & Mealybug Control (.02%) - Dragon
Spider Mite Control (.057%) - Dexol
Wasp & Hornet Spray (.284%) - Green Light
Whitefly & Mealybug Control (.057%) - Dexol
Whitefly & Mealybug Spray (.02%) - Dragon

Aphids, fungus gnats, leafhoppers, mealybugs, plant bugs, spider mites, spittle bugs, thrips, white lfeas, wasps, hornets, yellow jackets, bees, spiders, scales, whiteflies.

TETRACHLORVINPHOS

Livestock Dust (3%) – Southern Ag

Fowl mites, flies, cattle grubs, lice.

HERBICIDES

2,4-D

2,4-D Amine Weed Killer (49.3%) –Southern Ag
Dandelion Killer (13.3%) – Lilly Miller
Spot Weed Killer (.65%) – Rigo
Weed & Feed (1.01%) – Pace

Dandelions, pennywort, plaintain, wild onions, common broadleaf weeds.

2,4-D/2,4-DP

Brush Buster (32.15%) – Lawn & Garden
  Products
Poison Ivy & Brush Killer (32.15%) – Bonide
Super Brush Killer (32.15%) – Dragon

Alder, ash, aspen, birch, blackberry, black cherry, oak, black locust, boxelder, brambles, buck brush, Ceanothus, chamise, coffeeberry, current, elderberry, fir, gooseberry, green briar, gum, hemlock, honeysuckle, locust, manzanita, maple, multiflora roses, osage orange, palmetto, pine, poison ivy, poison oak, poplar, red elder, red elm, red maple, salmonberry, sand sagebrush, serviceberry, snowberry, spruce, sumac, sycamore, tulip poplar, wild cherry, wild grape, willow, winged elm and many other species.

2,4-D/DICAMBA

Brushkil (14.8%) – Bonide
Lawn Weed Killer (14.8%) – Bonide

Burdock, thistle, morninglory, ragweed, yarrow, golden rot, poison ivy, poison oak, sumac, brambles, hardy weeds, brush, stump treatment, ragweed, dandelion, chickweed, dock, plaintain, clover, ground ivy, chicory, lambsquarters, mustard, wild ester, wild carrot, wild onion, wild garlic, wild radish, oxalis, knotweed.

2,4-D/MCPP

Weed-B-Gon Ready Spray (11.68%) – Solaris
Weed-B-Gon RTU (.4%) – Solaris
Yard Basics Weed Killer for Lawns (.4%) – Solaris

Dandelion, clover, plaintain and many other broad leaf weeds, bur clover, carpetweed, chickweed, creeping charlie, curly dock, dandelion, English dairy, false dandelion, filaree, Florida pusley, heartleaf dry mary, henbit, oxalis, pennywort, plaintain, purslane, red sorrel, sheep sorrel, spurges, spurweed, thistles, toad flax, wild carrot, wild geranium, wild onion.

2,4-D/MCPP/2,4-DP

Dandelion & Broadleaf Weed Control
  Concentrate (13.66%) – UHS
Green Sweep Weed & Feed (6.85%) – Solaris
Lawn Food & Weed Control (.937%) – Pace
Lawn Weed Killer (13.66%) – Rigo
Triamine Weed & Feed (.937%) – UHS
Weed Whacker (13.66%) –
  Lawn & Garden Products
Weed Whacker Jet Spray(.97%) –
  Lawn & Garden Products
Weed Warrior (13.66%) – Agr Evo
Weed Warrior Spot Weed Killer (.97%) –
  Agr Evo
Weed-B-Gon Jet Weeder (.734%) – Solaris
Yard Basics Lawn Weed & Feed (6.85%) –
  Solaris

Loco weed, marsh elder, pepperweed, peppergrass, pokeweed, poor joe, poverty weed, prickly lettuce, primrose, puncture vine, Russian thistle, rushes, shepards purse, sowthistle, speedwell, spurge, stinkweed, sumac, sunflower, sweet clover, tar weed, toad flax, tumbleweed, velvetleaf, veronica, vetch, virginia creeper, wild aster, wild carrot, wild radish, willow, witchweed, wood sorrel, yarrow and many other broadleaf weeds, buckhorn, bur lock, butler cup, Canada thistle, carpetweed, chickweed, chicory, clovers, cocklebur, dock, dog fenel, English daisy, Florida pusley, ground ivy, heal all, henbit, jimsonweed, knotweed, lambsquarters, mallow, morninglory, mustard, oxalis, pigweed, plaintain, poison ivy, poison oak, purslane, ragweed, red clover, red sorrel, sheep sorrel, smartweed, speedwell, spurweed, spurge, wild aster, wild carrot, wild garlic, wild geranium, wild lettuce, wild onion, wood sorrel, yarrow, alder, field cress, thistles, blue lettuce, box elder, broomweed, carpet weed, ragweed, catnip, cockle, offerweed, croton, indigo, dogbane, elderberry, fleabane, flexweed, galensoga, goatsbeard, hemp, honeysuckle, horsetail, mallow, nettle, ironweed, jewelweed, jimsonweed.

2,4-D/MCPP/DICAMBA

Broadleaf Weed & Dandelion Control (14.99%)
  – Lilly Miller
Brush-No-More (42.54%) – PBI Gordon
Ezy Spray Weed & Feed (6.85%) – Dexol
Hose & Go Weed & Feed (9.04%) – Lilly Miller
Lawn Spot Weeder RTU (.946%) – Bonide
Lawn Weed Killer (12.09%) – Dragon
Lawn Weed Killer (14.95%) – Southern Ag
Lawn Weed Killer (14.99%) – Lilly Miller
Lawn Weed Killer RTU (.936%) – Dragon
Liquid Weed & Feed 15-2-3 (4.88%) – Bonide
Poison Ivy Poison Oak Killer (.896%) – Dexol
Poison Oak Poison Ivy Killer (.946%) – Dragon
Rapid Green Weed & Feed (.69%) – Lilly Miller
Ringer Supreme Lawn Fertilizer & Weed Control
  – Ringer
RTU Lawn Weed Killer (.046%) – Lilly Miller
RTU Spurge Oxalis & Dandelion Killer (.44%) –
  Lilly Miller
Spot Weeder (.946%) – Dexol
Spurge & Oxalis Killer (14.99%) – Lilly Miller

HERBICIDES

Spurge Oxalis & Dandelion Killer (14.99%) – Lilly Miller
Super Chickweed Killer (14.95%) – PBI Gordon
Super Rich Weed & Feed (.44%) – Lilly Miller
Trimec Lawn Weed Killer (12.09%) – PBI Gordon
Trimec Weed & Feed 24-4-8 (.703%) – PBI Gordon
Ultra Green Weed & Feed (.69%) – Lilly Miller
Weed & Feed 12-2-4 (.527% ai) – UHS
Weed & Feed Granules 17-4-4 (1.043%) – Bonide
Weed-B-Gon for Southern Lawns (14.95%) – Solaris
Weed-B-Gon Lawn Weed Killer 2 (14.95%) – Solaris
Weed-No-More Spot Weeder (.946%) –PBI Gordon
Weed-Out Lawn Weed Killer (12.09%) – Dexol
Wipe Out Broadleaf Weed Killer (9.71%) – Green Light
Wipe Out RTU (.946%) – Green Light

Brush & broadleaf weeds, kudyu, poison ivy, poison oak, ash, aspen, oak, brambles, willows, birches, black cherry, elms, gooseberry, honey locust, multiflora roses, short leaf pine, bedstraw, black medic, buck thorn, bull thistles, burdock, chicory, chick weed, clover, dandelion, dock, ground ivy, heal-all, henbit, knotweed, lambsquarters, lespedeza, mallow, morninglory, oxalis, peppergrass, pigweed, plaintain, purslane, ragweed, sheep sorrel, shepards purse, speedwell, spurge, thistle, wild carrot, wild garlic, wild lettuce, wild onion, yarrow, black medic, buckhorn, clover, dandelion, knotweed, wild lettuce.

2,4-D/MCPP/DICAMBA/MSMA

Crabgrass Plus Broadleaf Weed Killer (31.17%) – Dragon
Trimec Plus (31.17%) – PBI Gordon

Aster, bed straw, black medic, buckhorn, burdock, chicory, chickweed, clovers, crabgrass, dallisgrass, dandelion, docks, goosegrass, ground ivy, heal-all, henbit, knotweed, lambsquarters, lespedeza, mallow, morninglory, nut sedge, oxalis, peppergrass, pigweed, plaintain, poison oak, poison ivy, purslane, ragweed, sandbur, sheep sorrel, shepardspurse, speedwell, spurge, wild carrot, wild garlic, wild lettuce, wild onion, yarrow, yellow wood sorrel.

ATRAZINE

Atrazine 4L Herbicide (43%) – Southern Ag

Annual bluegrass, chickweed, crabgrass, cranesbill, cudweed, dichondra, Florida betony, henbit, knotweed, lespediaz, money wort, mustards, narrow leaf vetch, parsley-piert, sand spur, smut grass, spurge, spur weed, swine crest, wood sorrel, annual clovers.

BENEFIN

Balfin Granules (2%) – Southern Ag

Annual bluegrass, crabgrass, goosegrass, barnyardgrass, foxtails.

BENEFIN/ORYZALIN

Amaze Grass & Weed Preventer (2%) – Green Light

Bittercress, carpetweed, fiddleneck, chickweed, groundsel, purslane, rock purslane, Florida pusley, henbit, lambsquarters, London rocket, pigweed, knotweed, spurge, puncturevine, filaree, shepards purse, wood sorrel, oxalis, annual bluegrass, barnyard grass, bracharia, panicum, crabgrass, crowfoot grass, foxtails, sandbur, goosegrass, Italian ryegrass, johnsongrass, jungle rice, little barley, lovegrass, sprangletop, signalgrass, cupgrass, wild oats and witchgrass.

Partial control of sowthistle, mustards, nightshades, milkweed, mallow, ragweed, horseweed, lady thumb, morninglory, prickly lettuce, smartweed, spotted spurge, teaweed, velvetleaf and volunteer wheat.

BENEFIN/TRIFLURALIN

First Down Crabgrass Preventer (2%) – Green Light
Supreme Lawn Fertilizer & Crabgrass Preventer – Ringer
Ultragreen Crabgrass Control & Lawn Food (1.15%) – Lilly Miller

Annual bluegrass, crabgrass, goosegrass, crowfootgrass, silver crabgrass, barnyardgrass, foxtails, spurge (partial), oxalis (partial).

BENSULIDE

Betasan 3.6 Granules – Green Light

Crabgrass, pigweed, barnyardgrass, lambsquarters, goosegrass, annual bluegrass, shepardspurse, henbit, dead nettle.

CACODYLIC ACID

Liquid Edger (.62%) – PBI Gordon
Liquid Edger (.7%) – Green Light
Weed Ender (17.7%) – Lawn & Garden Products

Bermudagrass, johnsongrass, quackgrass, deep rooted perennials, brome, foxtail, wild oats, busclover, mallow, morninglory, annual & perennial weeds.

CAMA

Crabgrass Killer (8.4%) – Solaris
Crabgrass & Nutgrass Killer RTU (.5%) – Solaris

Crabgrass, dallisgrass, nutgrass.

HERBICIDES

COPPER SULFATE PENTAHYDRATE

Algae Attack (10.9%) – Lawn & Garden Products
Copper Sulfate Granules/Crystals (99%) – Southern Ag

Algae.

DCPA

Garden Turf & Ornamental Herbicide (5%) – Bonide
Garden Weed Preventer Granules (7.5%) – PBI Gordon
Turf & Ornamental Herbicide (75%) – Bonide
Vegetable Turf & Ornamental Herb. (75%) – Lawn & Garden Product
Weed Granules (5%) – Southern Ag

Crabgrass, purslane, lambsquarters, pigweed, chickweed, spotted spurge, lovegrass, carpet grass, witch grass, Florida pusley, foxtails, sandbur, dodder, barnyardgrass, goosegrass, panicum, ground cherry, annual bluegrass, johnsongrass, spurge.

DICLOBENIL

Casoron Granules (2%) – Lilly Miller
Casoron Granules (2%) – UHS
Casoron Granules (2%) – Solaris

Annual bluegrass, artemesia, bluegrass, bull thistle, camphorweed, Canada thistle, carpetweed, chickweed, citron melon, coffee weed, crabgrass, cudweed, curly dock, dandelion, dog fennel, evening primrose, false dandelion, fiddleneck, morninglory, Florida pusley, foxtails, gieskia, groundsel, henbit, horsetail, goosefoot, knotweed, lambsquarters, milkweed, natalgrass, peppergrass, pineapple weed, plantain, purslane, quackgrass, ragweed, dead nettle, pigweed, rosarypea, Russian knapweed, Russian thistle, shepardspurse, smartweed, Spanish needles, spurge, teaweed, Texas panicum, wild carrot, wild mustard, wild radish, yellow wood sorrel.

DIQUAT

Grass & Weed Killer (.23%) – Dragon
Grass & Weed Killer Concentrate (1.84%) – Dragon
Grass & Weed Killer RTU (.23%) – Bonide
Knock-Out Weed & Grass Killer (1.84%) – Lilly Miller
Liquid Edger (.23%) – Dexol
Poison Oak & Ivy Killer RTU (.23%) – Bonide
RTU Knock-Out Weed & Grass Killer (1.84%) – Lilly Miller
Weed & Grass Killer (.23%) – UHS
Weed & Grass Killer Concentrate (1.84%) – Dexol
Weed & Grass Killer RTU (.23%) – Dexol

Annual morninglory, foxtails, purslane, pigweed, bermudagrass and a wide variety of grasses and weeds.

EPTC

Eptam (2.3%) – Bonide
Eptam Weed & Grass Preventer (2.3%) – Green Light
Eptam Weed Control (2.3%) – Rigo
Pre-Emergent Weed & Grass Preventer (2.3%) – Dexol

Annual bluegrass, ryegrass, barnyardgrass, crabgrass, foxtails, goosegrass, johnsongrass, lovegrass, sandbur, volunteer graines, wild cane, wild oats, nightshade, chickweed, corn spurry, purslane, pigweed, ragweed, bermudagrass, mugwort, nutsedge, quackgrass.

FATTY ACIDS - POTASSIUM SALTS

Home Deck & Patio Moss & Algae Killer (2% and 40%) – Ringer
Lawn Moss Killer Concentrate (40%) – Ringer
Super Fast Weed & Grass Killer (3% and 18%) – Ringer

Annual and perennial weeds, moss, algae, lichens, liverworts, grime.

FERRIC SULFATE

Hose-N-Go Moss Out (35%) – Lilly Miller
Moss Out (35%) – Lilly Miller

Lawn moss.

FERROUS SULFATE

Lawn Food plus Moss Control (32%) – UHS
Moss Out Granules (95.4%) – Lilly Miller
Moss Out Lawn Granules (32%) – Lilly Miller
Moss Out Plus Fertilizer (32%) – Lilly Miller
Super Rich Lawn Food with Moss Control (11.2%) – Lilly Miller
Ultra Green Moss Control Lawn Food (32%) – Lilly Miller

Moss.

FLUAZIFOP-P-BUTYL

Grass-B-Gon (.48%) – Solaris
Grass-Out (1.7%) – Dexol

Annual bluegrass, bahiagrass, barnyardgrass, bentgrass, bermudagrass, centipede grass, crabgrass, dallis grass, downy brome, foxtails, goosegrass, johnsongrass, Kentucky bluegrass, nimblewill, orchardgrass, quackgrass, panicum, ryegrass, sandbur, St. Augustine grass, tall fescue.

HERBICIDES

GLUFOSINATE-AMMONIUM

Finale Concentrate (5.78%) – Agr Evo
Finale Ready to Use (1%) – Agr Evo
Finale Super Concentrate (11.33%) – Agr Evo

Annual sowthistle, bindweed, buffalobur, Canada thistle, chickweed, clover, cocklebur, curly dock, dandelion, dogbane, gromwill, filarea, fleabane, goldenrod, horsetail, jimsonweed, kachia, lambs quarter, leafy spurge, London rocket, malva, marestail, mush thistle, nettle, nightshade, penny cress, pigweed, plaintain, prickly lettuce, puslane, ragweed, Russia thistle, shepards purse, smartweed, tansy mustard, velvetleaf, vervain, copperleaf, aster, mustard wild onion, wild buckwheat, wild rose, wild turnip, wood sorrel, yellow rocket, annual bluegrass, bahiagrass, barley, barnyard grass, bermuda grass, carpet grass, crabgrass, cupgrass, Dallis grass, brome grasses, panicum, fescue, foxtails, goosegrass, guineagrass, johnsongrass, bluegrass, lovegrass, nutsedge, paragrass, quackgrass, ryegrass, sandbur, shatter cane, stink grass, torpedo grass, vasey grass, wheat, wild oats, wind grass, blackberry, pines, round leaf green briar, poison oak, poison ivy.

GLYPHOSATE

Fence & Yard Edger RTU (.96%) – Solaris
Kleenup Systemic Weed & Grass Killer (5%) – UHS
Kleeraway (7.5%) – Solaris
Roundup Concentrate (18%) – Solaris
Roundup RTU (.96%) – Solaris
Roundup Super Concentrate (41%) – Solaris
Roundup Tough Weed Formula (18%) – Solaris
Weed & Grass Killer (.96%) – Bonide
Weed & Grass Killer Kleenup (.5%) – UHS

Bahiagrass, bentgrass, bermudagrass, bindweed, blackberries, black medic, bluegrass, blue toad flax, brass buttons, broadleaf plaintain, brome grass, bur clover, Canada thistle, chickweed, groundsel, plaintain, crabgrass, beggarweed, creeping charlie, curly dock, dandelion, lovegrass, dog fennel, evening primrose, false dandelion, fennel, fescue, fiddleneck, filaree, Florida pusley, garden spurge, henbit, johnsongrass, knotweed, lambsquarters, little bitter cress, london rocket, maidencane, mallow, mayweed, nimblewill, oldenlamdia, orchard grass, oxalis, smartweed, pennywort, rye grass, poison ivy, poison oak, primrose, spurge, puncturevine, quack grass, ragweed, sandspur, shepards purse, smooth cats ear, spotted spurge, St. Augustine grass, sowthistle, tall fescue, tansy rag wort, white clover, white top, wild barley, yellow nutgrass, zoysia, creeping lantana, cud weed, foxtails, goosegrass, mayweed, prickly lettuce, pigweed, dead nettle, velvetleaf, wild geranium, bull thistle, centipede grass, hypericum, quickgrass, thistle, Virginia creeper, wild carrot, alder, ash, aspen, blackbeans, broom, buckwheat, ceanothis, chamise, cherry, coyote brush, dew berry, elderberry, elm, eucalyptus, hazel, hawthorne, honeysuckle, kudzu, locust, maple, oak, persimmon, poplar, raspberry, sage, sagebrush, salmonberry, sasafras, sumac, sweet gum, thimbleberry, tree tobacco, trumpet creeper, willow, wild rose.

GLYPHOSATE/ACIFLUORFEN

Grass & Weed Killer Kleen Up RTU (.62%) – UHS
Kleeraway Grass & Weed Killer (.62%) – Solaris
Yard Basics Weed & Grass Killer (.62%) – Solaris

Annual and perennial broadleaf weeds and grasses.

GLYPHOSATE/OXYFLUORFEN

Ground Clear Super Edger (.5%) – Solaris

Bahiagrass, bent grass, bermuda grass, bluegrass, bromegrass, crabgrass, lovegrass, fescue, johnsongrass, maiden cane, orchardgrass, ryegrass, quack grass, sand bur, St. Augustine grass, torpedo grass, white top, wild barley, nutgrass, zoysia, bind weed, balck medic, toad flax, brass buttons, plaintain, Canada thistle, chickweed, groundsel, beggarweed, curly dock, creeping charlie, dandelion, dog fennel, evening primrose, false dandelion, fennel, fiddleneck, filaree, Florida pusley, spurge, henbit, knotweed, lambs quarters, bitter cress, London rocket, mallow, mayweed, nimblewill, olden landia, oxalis, smartweed, penny wort, primrose, puncture vine, ragweed, shepards purse, smooth cats ear, sow thistle, ragwort, white clover, morninglory.

ISOXABEN

Protrait Broadleaf Weed Preventer (.38%) – Green Light

Bur weed, buttercup, wild celery, chickweed, fools parsley, Carolina geranium, henbit, knotweed, mallow, parsley, penny wort, pinapple weed, speedwill, spurge, corn spurry, wood sorrel, oxalis, (partial control of clovers, dandelions, black medic and plaintains).

MSMA

Crabgrass & Nutgrass Killer (16.6%) – Dragon
Crabgrass & Nutgrass Killer (16.8%) – PBI Gordon
Crabgrass Killer (13.4%) – Rigo
Monterey Weed-Hoe (47.8%) – Lawn & Garden Products
MSMA Crabgrass Killer (47.6%) – Bonide
MSMA Crabgrass Killer (48.3%) – Green Light

Crabgrass, nutgrass, dallis grass, sandbur, goosegrass, water grass, love grass, barnyard grass, lemon grass, chickweed, witch grass, carpet grass, cocklebur, nutsedge, ragweed, puncture vine, foxtails.

ORYZALIN

Hose-N-Go Weed & Grass Preventer (2.84%) – Lilly Miller

HERBICIDES

Surflan A.S. (40.4%) – Southern Ag
Weed & Grass Preventer (2.84%) – Lilly Miller
Weed Stopper (40.4%) –
 Lawn & Garden Products

Little barley, barnyard grass, annual bluegrass, crabgrass, crow foot grass, cupgrass, foxtails, goosegrass, johnson grass, jungle rice, lovegrass, wild oats, panicums, ryegrass, sandbur, spangle top, witchgrass, bitter cress, carpetweed, chickweed, fiddleneck, filaree, groundsel, henbit, knotweed, lambsquarters, pigweed, puncturevine, purslane, Florida pusley, London rocket, rock purslane, shepardspurse, spurge, yellow wood sorrel.

OXYFLUORFEN/IMAZAPYR

Ground Clear Triox (.78%) – Solaris

All vegetation.

PROMETON

Barren Vegetation Killer (1.5%) – Dexol
Com-Pleet (3.73%) – Green Light
Noxall Vegetation Killer (2.5%) – Lilly Miller
Total Vegetation Killer (3.75%) – Dragon
Triox (1.86%) – Solaris
Vegetation Killer (3.6%) – PBI Gordon

Grasses and undesirable weeds, brome grass, oat grass, goose grass, quack grass, puncture vine, golden rod, plaintain, bind weed, johnsongrass, wild carrot, morninglory, bermudagrass.

SETHOXYDIM

Poast (18%) – Lawn & Garden Products

Barnyardgrass, bermudagrass, broadleaf signal grass, crabgrass, panicum, foxtails, goosegrass, johnsongrass, jungle rice, lovegrass, orchardgrass, quackgrass, tall fescue, shattercane, wild cane, wirestem mukly, witch grass, woolly cupgrass.

SIDURON

Crabgrass Preventer & Weed Killer (2.75%) –
 Bonide

Crabgrass, foxtail, barnyardgrass.

SODIUM CHLORATE

Grass Weed & Vegetation Killer (18.5%) – Rigo

Annual and perennial weeds and grasses.

SODIUM CHLORATE/DIQUAT

Crack & Crevise Weed Killer (2.454%) – Dexol
Lawn Edging Liquid (2.454%) – Rigo
Weed & Grass Killer (2.454%) – Rigo

Woody perennials, poison oak, poison ivy, poison sumac, leafy spurge, bitter dock, golden rod, ragweed, milkweed, blueweed, crabgrass, bromegrass, chickweed, cocklebur, jimsonweed, lambsquarters, larkspur, prickly lettuce, shepards purse.

SODIUM CHLORATE/SODIUM META BORATE

Noxall Vegetation Killer (98%) – Lilly Miller

Annual weed, perennial weeds, johnson grass.

TRICLOPYR

Blackberry & Brush Killer (8%) – Lilly Miller
Brush Killer (8%) – Lilly Miller
Brush Killer (8.8%) – Dexol
Brush Killer (8.8%) – Southern Ag
Brush-B-Gon (8%) – Solaris
Brush-B-Gon 3 (.7%) – Solaris
Brush-B-Gon RTU (.7%) – Solaris

Alder, Arkansas rose, arrow wood, ash, aspen, beech, birch, blackberry, blackgum, box elder, California rose, cascara, ceanothus, cherry, chinquapin, choke cherry, cotton wood, Douglas fir, dogwood, elderberry, elm, hawthorne, hazel, honeysuckle, hornbean, horsetail rush, kudzu, locust, madrone, maples, mesquite, mimosa, mulberry, oaks, persimmon, pine, poison ivy, poison oak, poplar, raspberry, salmonberry, sassafras, scotch bloom, sumac, sweetbay, magnolia, sweet gum, sycamore, tan oak, tumbleberry, trumpet creeper, tulip poplar, vine maple, Virginia creeper, hemlock, wild grape, wild rose, willow, winged elm.

TRIFLURALIN

Rose Guard (.17%) – Rigo

Crabgrass, barnyardgrass, foxtails, johnson grass, goosegrass, annual bluegrass, stink grass, brome grass, jungle rice, sprangletop, pigweed, careless weed, lambsquarters, carpet weed, Russia thistle, kochia, purslane, Florida pusley, knotweed, stinging nettle, goosefoot, chickweed, puncture vine.

ZINC CHORIDE

Moss-Kil (29.6%) – Lilly Miller
RTU Moss-Kil (6.2%) – Lilly Miller

Moss.

ZINC SULFATE

Moss Kill Granules (99%)
Moss Kill Roof Strip (99%)

Moss.

FUNGICIDES

CALCIUM

Foli-Cal (10%) – Lawn & Garden Products
Tomato Blossom End Pest Spray (90%) – Dragon

Bitter pit, blossom end-rot, black heart, internal brown spot, cork spot, internal browning.

CALCIUM POLYSULFIDE

Dormant Disease Control (26%) – Solaris
Lime Sulfur Solution (29%) – Dragon
Lime Sulfur Spray (29%) – PBI Gordon
Polysul (28.7%) – Lilly Miller
Sulf-R-Spray (28.7%) – Lilly Miller

Scales, mites, powdery mildew, rust, scab, blotch, peach leaf curl, brown rot, twig borers, anthracnose, larch case bearer, leaf blotch, maple gall, boxwood canker, nectria canker, black spot, shot hole, red berry trouble, leaf spot.

CAPTAN

Wettable Captan (50%) – Dragon
Captan 50% WP - Bonide
Captan (5.1%) – Lilly Miller
Captan Fungicide (50%) – Southern Ag

Scab, fruit spot, bitter rot, black rot, Botrytis, blossom end rot, sooty blotch, fly speck, black pox, Botryosphaeria rot, brown rot, leaf spot, Botrytis rot, downy mildew, scab, brown rot, gray mold, damping off, brown patch, melting out, seedling blight, brown spot, black spot, petal blight, Septoria leaf spot.

CAPTAN/MALATHION/METHOXYCHLOR

Home Orchard Spray (37.5%) – Solaris

INSECTS – Aphids, curculio, leafhopper, leaf roller, pear slug, mites, tent caterpillar thrips, apple maggot, canker worm, codling moth, flea beetle, Japanese beetle, Oriental fruit moth, plant bugs, peach tree borer, peach twig borer, fruit moth, pear psylla, spittle bug, strawberry leaf beetle, strawberry weevil, grape berry moth, rose chafer.

DISEASES – Black rot, dead-arm, Botrytis rot, brown rot, scab, bitter rot, black rot, Botryosphaeria, black pox, brooks fruit spot, fly speck, scab, brown rot.

CAPTAN/MALATHION/METHOXYCHLOR/CARBARYL

Complete Fruit Tree Spray (30.3%) – Bonide
Fruit Tree Spray (15.1%) – Bonide
Liquid Fruit Tree Spray (30.3%) – PBI Gordon
Rose & Flower Dust (15.5%) – Dragon
Rose & Flower Insect & Disease Concentrate (30.3%) – Dragon
Rose & Flower Spray or Dust (15.5%) – Bonide
Fruit Tree Spray (30.3%) – Dragon

INSECTS – Strawberry weevil, aphids, mites, lygus, fleabettles, omnivorous leaf tier, spittlebug, Japanese beetles, crickets, apple maggot, bagworm, cutworm, bud moth, cherry fruit fly, codling moth, plum curculio, leaf roller, gypsy moth, leaf hopper, peach tree borer, mealy bugs, mites, oriental fruit moth, pear slugs, psylla, scales, tent caterpillar, leaf miners, yellow neck caterpillar.

DISEASES – Botrytis rot, leaf spots, bitter rot, black rot, blossom blight, Botryasphaeria rot, Botrytis blossom end rot, Brooks fruit rot, brown rot, Coryreum blight, downy mildew, fly speck, frog eye, fruit rot, scab, sooty blotch.

CAPTAN/MALATHION/METHOXYCHLOR/SULFUR

Fruit Spray Concentrate (9.46%) – Southern Ag

INSECTS – Aphids, mites, bud moth, codling moth, leaf rollers, apple maggot, Japanese beetle, tent, caterpillars, fruit flies, tree rollers, fruit worm, rose chafer, canker worm, mealy bugs, leaf hoppers, berry moth, leaf skeletonizer, Oriental fruit moth, strawberry weevil, flea beetles, spittle bugs.

DISEASES – Botrytis rot, leaf spots, powdery mildew, brown rot, Rhizophus, scab, rust, downy mildew, black rot, rust, bitter rot, white rot, sooty blotch, frogeye leaf spot.

CHLOROTHALONIL

Daconil 2787 (12.9%) – Dragon
Daconil 2787 (29.6%) – Solaris
Daconil Lawn & Garden Fungicide (12.5%) – Lilly Miller
Disease Control (12.5%) – Lilly Miller
Fungicide (12.5%) – Lilly Miller
Lawn & Garden Fungicide (12.5%) – Rigo
Lawn Ornamental & Vegetable Flowable Fungicide (12.5%) – Southern Ag
Lawn Ornamental & Vegetable Fungicide (75%) – Southern Ag
Monterey Bravo Flowable Fungicide (40.4%) – Lawn & Garden Products
Multi-Purpose Fungicide (12.5%) – PBI Gordon

Copper spot, Curvularia, fading out, leaf spot, dollar spot, Sclerotinia, gray leaf spot, Helminthosporium, melting out, large brown patch, Rhizoctonia, red thread, stem rust, Alternaria leaf spot, anthracnose, gray snow mold, Typhula, Cercospora, Ovrilinia petal blight, leaf blotch, stem rot, Botrytis blossom blight, scab, gray mold, ray blight, Fusarium leaf spot, brown rot, Monilinia, cedar apple rust, Septoria, leaf & flower spot, Rhizoctoria web blight, Ascochyta blight, Taphrina blister, tan leaf spot, Cipolaris leaf spot, Phytophthora blight and die back, crown rot, black

FUNGICIDES

spot, Fabraea leaf spot, Marssonia leaf spot, leaf blight, Cephalosporium leaf spot, powdery mildew, Swiss needle cast, Rhabdocline needle cast, Sirococcus tip blight, Scheroderris canker, pine brown spot, Rhizosphaera needle rot, twig blight, shot hole, peach leaf curl, downy mildew, gray mold blight, ring spot, early blight, late blight, basal stock rot, pink rot, cherry leaf spot, gummy stem blight, target leaf spot, Botrytis leaf blotch, purple blotch, stem end rot, bottom rot, black mold rot.

COPPER - COPPER AMMONIA COMPLEX

Kop-R-Spray (8%) – Lilly Miller
Liqui Cop (8%) – Lawn & Garden Products

Alternaria, angular leaf spot, Anthracnose, bacterial blight, bacterial leaf spot, black canker, black rot, black spot, blossom blight, Botrytis blight, brown rot, Coryneum blight, dieback, downy mildew, early and late blight, false smut, halo blight, leaf blight, leaf blister, leaf blotch, leaf gall, leaf scorch, leaf spots, lichens, nectria canker, needle cast, peach leaf curl, powdery mildew, scab, rusts, shot hole, stem canker, twig blights, walnut blight.

COPPER - BASIC COPPER SULFATE

Copper Fungicide (90%) – Lilly Miller
Copper Spray or Dust (7%) – Bonide
Microcop (90%) – Lilly Miller
Tomato Dust (13.3%) – Southern Ag

Downy mildew, leaf spot, early & late blight, powdery mildew, gummy stem blight, snthracnose, scab, bacterial spot, leaf mold, rust, leaf scorch, bitter rot, black rot, blotch, brown rot, blossom blight, shot hole, leaf blister, Botrytis blight, stem canker, brown needle blight, leaf curl, anthracnose, perennial canker, Coryneium blight, damping off, bacterial blight, late blight, galls, dieback, scab, powdery mildew, bud & twig blight.

COPPER - BORDEAUX MIXTURE

Bordeaux Mix (12.75%) – Dragon
Bordeaux Mixture (12.75%) – Dexol
Bordeaux Mixture (13.3%) – PBI Gordon
Bordeaux Powder (12.75%) – Rigo

Scab, black rot, bitter rot, blotch, fire blight, bacterial leaf spot, brown rot, peach leaf curl, pecan scab, powdery mildew, early blight, late blight, leaf spot, melanose, Septoria leaf spot, anthracnose, elm leaf curl, needle cast, twig blight, Sphaeropsis blight, coryneum blight, cedar apple rust, leaf blight, nectria canker, dieback, Botrytis blight, gray mold, leaf spot, leaf blotch.

COPPER/CARBARYL

Copper Dragon (9%) – Dragon
Copper Dragon Tomato & Vegetable Dust (9%) – Dragon

INSECTS – Bean beetle, leaf hoppers, flea beetles, Japanese beetle, velvetbean beetle, lygus, stinkbug, corn earworm, Colorado potato beetle, army worm, hornworm, European corn borer, stink bugs, lace bugs, pickle worm, melonworm, cucumber beetle, squash bugs, Harlequin bug, spittlebugs, asparagus beetle, cabbage caterpillar, bagworm, tent caterpillar, grape berry moth, June beetle, scale, leaf rollers, strawberry weevil, tarnished plant bug.

DISEASES – Leaf spot, early & late blight, Cercospora leaf spot, Anthracnose, bacterial wilt, downy mildew, scab, Alternaria leaf spot, powdery mildew.

COPPER - FATTY & ROSIN ACIDS

Copper Fungicide (4%) – Dragon
Liquid Copper Fungicide (4%) – Rigo
Liquid Copper Fungicide (48%) – Southern Ag

Melanose, greasy spot, downy mildew, powdery mildew, anthracnose, needle blight, bacterial blight, alternaria blight, early blight, Septoria leaf spot, leaf spot, bacterial spot and speck, bacterial canker, Cercospora leaf spot, brown rot, blossom blight, leaf curl, shot hole, web blotch, bacterial blight, cane spot, yellow rust, alternaria blight, late blight, angular leaf spot, bacterial soft rot, grey mold neck rot, botryis blight, leaf gall, Phytophthora dieback, fireblight, scabs, needle blight, tip blight, cedar apple rust.

FERBAM

Ferbam Wettable Fungicide (76%) – Dragon

Scab, cedar rust, blotch, rust, black rot, sooty blotch, fly speck, brooks spot, bitter rot, black spot, leaf spot, Botrytis blight, leaf curl.

MANCOZEB

Broad Spectrum Mancozeb Fungicide (80%) – Green Light
Dithane M-45 (80%) – Southern Ag
Mancozeb Disease Control (37%) – Dragon
Mancozeb Plant Fungicide (80%) – Bonide

Asternaria leaf spot, anthracnose, Botrytis leaf blight, Cercospora leaf spot, rust, Crown rot, downy mildew, early blight, Fusarium decay, gray leaf spot, gummy stem blight, Helmuthosporium leaf blight, late blight, leaf mold, neck rot, purple blotch, scab, steptoris leaf spot, smut, black spot, blight, Botrytis petal spot, brown spot, canker cedar apple rust, Cercospora frog eye, clylendrocladium, rot, leaf blotch, petal blight, purple spot, smoulder, spadex rot.

FUNGICIDES

MANEB

Maneb Tomato & Vegetable Fungicide (80%) – PBI Gordon

Alternaria leaf spot, downy mildew, gummy stem blight, Helminthosporium leaf blight, anthracnose, fruit rot, Botrytis leaf blight, purple blotch, early blight, late blight, angular leaf spot, Septoria leaf spot, gray leaf mold, gray mold, black rot, bunch rot, Botrytis blight, Phytophthora blight, black spot, rust, Curvularis, stemphylium, Sclerotinia, dollar spot, melting out, brown patch, Rhizoctoria.

MANEB/CARBARYL

Tomato - Potato Dust (7.8%) – Bonide

INSECTS – Fruit worm, army worm, horn worm, European corn borer, stink bug, lace bugs, lygus, Colorado potato beetle, flea beetle, flea hopper, pickleworm, melonworm, cucumber beetle, squash bug, harlequin bug, imported cabbageworm, loopers, corn earworm, corn rootworm, Japanese beetle, sap beetle, budworm, Mexican bean beetle, bean leaf beetle, cutworm, velvetbean caterpillar, tarnished plant bug, spittle bug.

DISEASES – Anthracnose, early & late blight, downy mildew, Septoria leaf spot, gray leaf spot, gray mold, fruit rot, gummy stem blight, alternaria leaf spot, Helminthosporium leaf blight, rust.

PCNB

Fungicide (50%) – Lilly Miller

Brown patch, petal blight, flower blight, crown rot, black rot, stem rot, damping off.

SULFUR

Flower Fruit & Vegetable Garden Fungicide (.4% & 12%) – Ringer
Wettable or Dusting Garden Sulfer (90%) – Dragon
Wettable Dusting Sulfur (90%) – Agr Evo
Liquid Sulfur (52%) – Bonide
Sulfur Plant Fungicide (95%) – Bonide
Sulfur Dust (90%) – Lilly Miller
Wettable Dusting Sulfur (90%) – Green Light
Wettable or Dusting Sulfur (90%) – Southern Ag

Powdery mildew, black spot, rust, leaf spot, brown rot, red spiders, scab, frog eye, black rot, sooty blotch, stem blight, scales, peach canker, shot hole.

SULFUR/MALATHION/CARBARYL

Rose & Flower Dust (40%) – Lilly Miller
Tomato & Vegetable Dust (40%) – Lilly Miller

INSECTS – Aphids, mites, hornworm, fruit worm, leaf hopper, loopers, imported cabbageworm, armyworm, flea beetles, ants, box elder bug, cutworm, earwigs, Japanese beetle, June beetle, leaf rollers, mealy bugs, tent caterpillars, thrips, whiteflies.

DISEASES – Powdery mildew, rust.

THIOPHANATE-METHYL

Rose & Ornamental Fungicide (50%) – Rigo
Systemic Fungicide 3336 WP (50%) – Dragon
Systemic Fungicide Disease Control (50%) – Green Light
Thiomyl (50%) – Southern Ag

Stem, crown & root rot, cylindrocladium rot, bulb rot, black spot, powdery mildew, Botrytis flower blight, scab, leaf blight, leaf spot, anthracnose, Botrytis blight, bulb rot, copper spot, crown rot, dollar spot, Fusarium blight, Fusarium patch, large brown patch, leaf spot, necrotic rug spot, ovulinia blight, powdery mildew, red thread, root rot, rust, stem blight, stripe smut, summer patch, tip blight, pink snow mold, cylindrocladium rot.

THIRAM/METHOXYCHLOR

Bulb Dust (15%) – Bonide
Bulb Dust (15%) – Lilly Miller

Basal rot & decay, storage rot, thrips.

TRIADIMEFON

Bayleton 25% Fungicide – Lawn & Garden Products
Bayleton Systemic Fungicide (25%) – Bonide
Fung-Away Systemic Fungicide (.88%) – Green Light
Fung-Away Systemic Lawn Fungicide Spray (.88%) – Green Light
Fung-Away Systemic Turf Fungicide (.5%) – Green Light
Procide (.88%) – Agr Evo
Procide G (.5%) – Agr Evo
Systemic Fungicide for Turf & Ornamental (.89%)
Turf Fungicide Granules (.5%) – Southern Ag

Dollar spot, copper spot, powdery mildew, red thread, rusts, Rhizoctonia blight,m anthracnose, Southern blight, stripe smut, Fusarium blight, summer patch, gray snow mold, Typhus blight, pink snow mold, Fusarium patch, Fusarium blight, cedar apple rust, black rot, flower blight, leaf blight, leaf spots, tip blight.

TRIFORINE

Funginex (6.5%) – Solaris

Black spot, powdery mildew, rusts, leaf spot, petal blight.

FUNGICIDES

TRIFORINE/ACEPHATE/HEXAKIS

Orthenex Insect & Disease Control (8%) – Solaris

INSECTS – Aphids, thrips, lace bugs, leaf hoppers, budworms, leaf miners, spittlebugs, mites.

DISEASES – Black spot, rust, powdery mildew.

TRIFORINE/ACEPHATE/RESMETHRIN

Orthenex Insect & Disease Control I (.464%) – Solaris

INSECTS – Aphids, whiteflies, mealy bugs, bud worms, mites, leaf miners, thrips, scales, armyworms, bagworms, diabrotica, lace bugs, leaf hoppers, leaf miners, leaf tiers, salt marsh, caterpillar, olerander caterpillars, rose midge.

DISEASES – Black spot, rust, powdery mildew.

MISCELLANEOUS CHEMICALS

GROWTH REGULATORS

CARBARYL

Sevin Liquid Insecticide (11.7%) – UHS

Reduces fruit production on apples and crabapples.

CYTOKININS

Berry & Fruit Set Spray (.001%) – Bonide
Blossom Set for Tomatoes & Vegetables (.000008%) – Dexol
Tomato & Blossom Set Spray (.000008%) – Bonide
Tomato & Vegetable Bloom Set (.000008%) – Bonide

Promotes bloom set, increases yields and speeds harvest on tomatoes, beans, cucumbers, eggplant, melons, okra, peppers, strawberries, grapes and other plants.

ETHEPHON

Floral Fruit Eliminator (.33%) – Lawn & Garden Products

Control nusiance fruit on apples, crabapples, carobs cottonwood, elm, flowering pear, horse chestnut, maple, oak, olive, pine, sour orange, sweetgum (liquid amber) and sycamore. Controls mistletoe in deciduous trees and conifers.

IBA (Indole Butyric Acid)

Rootone Rooting Powder (.1%) – Bonide

Promotes healthier roots from cuttings.

NAA/THIRAM

Rootone (4.64%) – Dragon
Rootone (4.64%) – Lilly Miller
Rootone Brand F (4.24%) – Dexol
Rootone Brand F (4.24%) – Green Light

Promotes healthier rootings from cuttings plus giving fungicidal activity.

P. CHLOROPHENOXYACETIC ACID

Tomato Bloom Spray II (.005%) – Green Light
Tomato Plus (.005%) – Lilly Miller

Helps tomatoes set blossoms and ripen 1-3 weeks early.

MOLLUSCICIDES

METALDEHYDE

Bug-Geta (3.25%) – Solaris
Deadline (4%) – Pace
Deadline Bullets (4%) – Pace
Hose-N-Go Snail & Slug Killer (10.5%) – Lilly Miller
Slug & Snail Bait (2%) – Lilly Miller
Slug & Snail Line (4%) – Lilly Miller
Slug-N-Snail Beater RTU (4%) – Bonide
Slug-N-Snail Spray (10.5%) – Lilly Miller
Snail & Slug Killer Pellets (2.75%) – Dragon
Snail & Slug Pellets (2%) – Lilly Miller
Snail & Slug Pellets (2%) – Lilly Miller

Controls snails and slugs.

METALDEHYDE/CARBARYL

Bait Pellets (7%) – Southern Ag
Bug & Snail Bait (5%) – Green Light
Bug-Geta Plus (7%) – Solaris
Go-West Meal (7%) – Lilly Miller
Slug, Snail & Insect Killer Bait (7%) – Lilly Miller
Slug, Snail & Sowbug Bait (7%) – Bonide
Slug-N-Snail Granules (8%) – Lilly Miller

Snails, slugs, ants, earwigs, armyworms, cutworm, sowbugs, millipedes, crickets, grasshoppers.

MISCELLANEOUS CHEMICALS

REPELLENTS

FATTY ACIDS - AMMONIUM SOAPS

Hinder (13.8%) – Pace

Repells deer and rabbits on annual and perennial flowers, nursery, stock, ornamental trees and shrubs, apples and pear trees, home gardens.

METHYL NONYL KETONE

Dog & Cat Repellent – Dexol
Scat Cat Repellent – Dragon

Repells dogs and cats in flowers, shrubs, trees and lawn areas.

NAPTHALENE

Mosquito Beater (5%) – Bonide
Rabbit & Dog Repellent (15%) – Dragon

Repells rabbits and dogs on ornamentals, shrubs, trees, flowers, garbage cans, barns and other areas. Also repels mosquitos.

POTASSIUM NITRATE

Stump Remover – Dexol
Stump Remover – Lilly Miller

Speeds up decomposition of cut stumps.

THIRAM

Rabbit - Deer Repellent & Bulb Saver (11%) – Bonide

Repells deer, rabbits, mice, moles, dogs, cats and other animals from desired shrubs and trees while they are dormant. Protects bulbs from moles and water loss.

RODENTICIDES

BROMADIOLONE

Rat & Mouse Bait (.005%) – Green Light

Controls mice and rats.

CHOLECALCIFEROL (Vitamin D)

Quintox (.075%) – Lawn & Garden Products

Controls mice and rats.

DIPHACINONE

Rat & Mouse Bar (.005%) – Green Light

Controls mice and rats.

POTASSIUM NITRATE/SULFUR

Gopher Gasser – Dexol

Controls gophers, moles and ground squirrels.

STRYCHNINE

Gopher Mix (.5%) – Lilly Miller

Controls gophers.

ZINC PHOSPHIDE

Gopher Killer Pellets (2%) – Dexol
Mole & Gopher Killer (2%) – PBI Gordon
Mole Killer Pellets (2%) – Dexol
Mole Killer Pellets (2%) – Dragon
Moletox II (2%) – Bonide

Controls moles and gophers.

A Guide To
Lawn, Garden & Home
Pest Control Products

Section III   **Useage Of Products**

Insecticides	**40**
Herbicides	**72**
Fungicides	**90**

USAGE OF PRODUCTS

INSECTICIDES

ABELIA
Systemic Rose & Flower Food

ACACIA
Spider Mite & Mealybug Control
Whitefly & Mealybug Spray

AFRICAN VIOLET
Aphid Mite & Whitefly Killer
Aphid Mite & Whitefly Killer II
Bug Off Rose & Flower Spray
Flower Garden Insecticidal Soap
Fruit & Veg Insect Spray
Garden Spray
Gentle Care House Plant Spray
Home Patrol Insect Killer
House Plant & Garden
    Insect Spray
Insecticidal Soap Multi Purpose
    Insect Killer
Japanese Beetle Killer
Mite Beater
Natural Pyrethrin
Plant Insect Control
Rose & Floral Insect Control
Rose & floral Insect Killer
Roses & Flower Insect Spray
Rose & Flower Insect Killer
Rose & Flower Insect Killer II
Rose, Flower & Ornamental
    Insect Spray
Rose & Garden Insect Fogger
Spider Mite Control
Tomato & Veg Insect Killer
Vegetable Insect Control
Vegetable Insect Killer
Whitefly & Mealybug Control

AGERATUM
2% Systemic Granules
Intercept H&G
Intercept RFO
Intercept Vegetable & Garden
    Spray
Permethrin T/O
Spider Mite & Meelybug Control
Systemic Granules
Systemic House Plant Insect Cotrol
Whitefly & Meelybug Spray

ALMOND
Bioneem
Diazinon 25%
Diazinon 25% EC
Dormant & Summer Oil Spray
Dursban Plus
Horticultural Spray Oil
Intercept Vegetable & Garden
    Spray
Japanese Beetle Repellent
Permethrin T/O

Saf-T-Side
Sevin Liquid
Sevin Spray/Liquid
Spray Oil
Summer & Dormant Oil
Sun Spray Ultra Fine
Worm Ender

ALMOND (flowering)
Intercept H&G

ALUMINUM PLANT
Aphid Mite Whitefly Killer II
Intercept RFO
Spider mite Control
Spider Mite & Meelybug Control
Whitefly & Meelybug Control
Whitefly & Meelybug Spray

AMARANTH
Intercept RFO

ANDROMEDA
Horticultural Spray Oil
Rose & Florel Spray Bomb
Sevin 5 Dust
Sevin 10 Dust

ANISE
Bioneem
Japanese Beetle Repellent

APPLES
25% Methoxychlor Spray
50% Malathion
Bioneem
Bug Buster-O
Diazinon Dust
Diazinon 12 1/2% E
Dormant Oil Spray
Dormant & Summer Oil Spray
Dormant Spray & Summer Spray
Dursban Plus
Fruit & Berry Insect Spray
Garden Insect Spray
Horticultural Spray Oil
Imidan
Japanese Beetle Repellent
Kelthane
Lindane Borer Spray
Liquid Flowable Sevin
Malathion 50%
Malathion 50 Plus
Methoxychlor 25%
Organic Greenhouse
    House & Veg. Spray
Permithrin T/O
Roses & Flower Insect Spray
Saf-T-Side
Scale Away
Sevin 50% WP
Sevin 50 WP

Sevin 50 Wettable
Sevin Liquid
Sevin Dura Spray
Soluable Oil Spray
Spray Oil
Summer & Dormant Oil
Sun Spray Ultra Fine
Thiodan Insect Spray
Thiodan Vegetable &
    Ornamental Dust
Volck Oil Spray
Worm Ender

APRICOTS
25% Diazinon Insect Spray
25% Methoxychlor Insect Spray
50% Malathion
Bioneem
Bug-B-Gon
Bug Buster-O
Diazinon 25%
Diazinon 25% EC
Diazinon Plus
Dormant & Summer Oil Spray
Garden Insect Spray
Horticultural Spray Oil
Imidan
Japanese Beetle Repellent
Liquid Flowable Sevin
Malathion 50%
Malathion 50 Plus
Methoxychlor 25%
Saf-T-Side
Scale Away
Sevin 50% WP
Sevin 50 WP
Sevin Dura Spray
Sevin Liquid
Spray Oil
Summer & Dormant Oil
Sun Spray Ultra Fine
Thiodan Insect Spray
Volck Oil Spray
Worm Ender

ARALIA
Aphid Mite & Whitefly Killer II
Intercept RFO
Spider Mite Control
Systemic Rose & Flower Food
Whitefly & Mealybug Control

ARALIA (Fatsia)
Aphid Mite & Whitefly Killer II
Intercept H&G
Spider Mite Control
Whitefly & Mealybug Control

ARBORVITAE
25% Diazinon Insect Spray
Bagworm & Mite Spray
Diazinon 4E

40

INSECTICIDES

Diazinon 25%
Diazinon 25% EC
Diazinon Spray
Horticultural Spray Oil
Imidan
Intercept H&G
Intercept RFO
Liquid Sevin
Rose & Florel Spray Bomb
Sevin 5 Dust
Sevin 10 Dust

ARDISIA

Intercept Vegetable & Garden Spray

ARIZONA CYPRESS

Permethrin T/O

ARROWHEAD VINE

Intercept RFO
Spider Mite & Meelybug Control
Whitefly & Meelybug Spray

ARTICHOKES

Bioneem
Bug Buster
Bug Buster-O
Garden Insect Spray
Japanese Beetle Repellent
Thiodan Insect Spray
Thiodan Vegetable & Ornamental Dust
Worm Ender

ARUGOLA

Worm Ender

ASH

Borer Killer II
Borer Miner Killer 5
Imidan
Sevin 5 Dust
Sevin 10 Dust

ASPARAGUS

Bioneem
Blue Dragon Garden Dust
Bug Buster-O
Cutworm Earwig & Sowbug Bait
Earwig Sowbug & Grasshopper Bait
Fruit & Veg Insect Spray
Garden Spray
Grasshopper Earwig & Sowbug Bait
House Plant & Garden Insect Spray
Intercept Vegetable & Garden Spray
Japanese Beetle Killer
Japanese Beetle Repellent

Malathion 50%
Natural Pyrethrin
Organic Greenhouse House & Veg. Spray
Permethrin T/O
Roses & Flower Insect Spray
Saf-T-Side
Sevin Granules
Sevin Liquid
Sevin Spray/Liquid
Sun Spray Ultra Fine
Tomato & Vegetable Insect Killer
Tomato & Vegetable Insect Spray
Tomato & Vegetable Fogger
Tomato & Veg. RTU
Tomato Pepper Veg. Spray RTU
Vegetable Insect Control
Vegetable Insect Killer
Yard & Garden Insect Killer
Worm Ender

ASPEN

Borer Miner Killer 5

ASTERS

5% Malathion Dust
Bug Off Rose & Flower Spray
Garden Spray
Gentle Care House Plant Spray
Home Patrol Insect Killer
Intercept RFO
Intercept Vegetable & Garden Spray
Japanese Beetle Killer
Mite Beater
Natural Pyrethrin
Permethrin T/O
Plant Insect Control
Rose & Floral Insect Control
Rose & Floral Insect Killer
Rose & Flower Insect Killer II
Rose & Flower Spray
Rose & Flower Dust
Rose, Flower & Ornamental Insect Spray
Rose & Garden Insect Fogger
Sevin 5 Dust
Sevin 10 Dust
Systemic House Plant Insect Control
Tomato & Vegetable Insect Killer
Vegetable Insect Control
Vegetable Insect Killer
Whitefly & Meelybug Spray
Yard & Garden Insect Killer

AVOCADO

Malathion 50%
Malathion-Oil
Saf-T-Side
Soluable Oil Spray
Worm Ender

AZALEA

2% Systemic Granules
5% Diazinon Dust
5% Malathon Dusst
25% Diazinon
50% Malathion
Ant Killer Powder
Bagworm & Mite Spray
Blue Dragon Garden Dust
Bug-B-Gon
Bug Off Rose & Flower Spray
Cygon 2E
Diazinon 4E
Diazinon 12 1/2% E
Diazinon 25%
Diazinon 25% EC
Diazinon Plus
Diazinon Spray
Dormant & Summer Oil Spray
Dormant Spray & Summer Spray
Feed & Shield Systemic Rose Care
Fruit & Veg Insect Spray
Garden Spray
Gentle Care House Plant Care
Home Patrol Insect Killer
Horticultural Spray Oil
House Plant & Garden Insect Spray
Intercept H&G
Intercept Vegetable & Garden Spray
Japanese Beetle Killer
Lindane Borer Spray
Liquid Sevin
Malathion 50 Plus
Mite Beater
Mite & Insect Spray
Multi Purpose Insect Killer
Natural Pyrethrin
Permethrin T/O
Rose & Floral Insect Control
Rose & Floral Insect Killer
Rose & Floral Spray Bomb
Rose & Flower Insect Killer
Rose & Flower Insect Killer II
Rose & Flower Insect Spray
Rose & Flower Spray
Rose, Flower & Ornamental Insect Spray
Rose & Garden Insect Fogger
Saf-T-Side
Sevin 50% WP
Sevin 5% Dust
Sevin 50 Wettable
Sevin 50WP
Soluable Oil Spray
Summer & Dormant Oil
Sun Spray Ultra Fine
Systemic Granules
Systemic House Plant Insect Control

USAGE OF PRODUCTS

USAGE OF PRODUCTS

INSECTICIDES

Systemic Rose & Flower Food
Tomato & Vegetable Insect Killer
Vegetable Insect Control
Vegetable Insect Killer
Volck Oil Spray
Whitefly & Mealybug Spray
Yard & Garden Insect Killer

AZARINA
Intercept H&G

BABY'S BREATH
Intercept Vegetable & Garden Spray

BABY'S TEARS
Intercept RFO
Spider Mite & Meelybug Control
Whitefly & Meelybug Spray

BAMBOO
Sevin 5 Dust
Sevin 10 Dust

BANANA
Saf-T-Side
Worm Ender

BARBERRY
Intercept H&G

BASIL
Bioneem
Flower Garden Insecticidal Soap
Insecticidal Soap Multi Purpose Insect Killer
Japanese Beetle Repellent

BEANS
5% Diazinon Granules
5% Malathion Dust
5% Sevin Garden Dust
25% Diazinon
25% Methoxychlor Insect Spray
50% Malathion
Bacillus thuringiensis
Bioneem
Blue Dragon Garden Dust
Bug-B-Gon
Bug Buster
Cutworm Earwig & Sowbug Bait
Cygon 2E
Diazinon 2% Granules
Diazinon 5% Granules
Diazinon 5% G
Diazinon 12 1/2% E
Diazinon 25%
Diazinon 25% EC
Diazinon 5G
Diazinon Dust
Diazinon Granules
Diazinon Plus

Diazinon Soil & Turf Insect Control
Diazinon Spray
Dipel
Earwig Sowbug & Grasshopper Bait
Flower Garden Insecticidal Soap
Fruit & Berry Insect Spray
Fruit & Veg Insect Spray
Garden Insect Spray
Garden Soil Insecticide
Garden Spray
Grasshopper Earwig & Sowbug Bait
House Plant & Garden Insect Spray
Japanese Beetle Killer
Japanese Beetle Repellent
Kelthane
Liquid Flowable Sevin
Liquid Sevin
Malathion 50%
Malathion 50 Plus
Many Purpose Insect Killer
Methoxychlor 25%
Mole Cricket Killer
Natural Pyrethrin
Organic Greenhouse House & Veg. Spray
Red Spider Spray
Roses & Flower Insect Spray
Saf-T-Side
Sevin 5 Dust
Sevin 5% Dust
Sevin 10% Dust
Sevin 10 Dust
Sevin 50% WP
Sevin 50 Wettable
Sevin Garden Dust
Sevin Granules
Sevin Liquid
Sevin Dura-Spray
Sevin Spray/Liquid
Sun Spray Ultra
Systemic Granules
Thiodan Insect Spray
Thiodan Vegetable & Ornamental Dust
Tomato & Vegetable Insect Killer
Tomato & Vegetable Dust
Tomato & Vegetable Fogger
Tomato & Vegetable Insect Killer
Tomato & Veg. RTU
Tomato Pepper Veg. Spray RTU
Thuracide
Vegetable Insect Control
Vegetable Insect Killer
Worm Ender
Yard & Garden Insect Killer

BEDDING PLANTS (general)
Bioneem
Bug Buster-O

Japanese Beetle Repellent
Rose & Flower Care
Saf-T-Side
Sun Spray Ultra Fine
Systemic Rose Care
Systemic Rose Shrub & Flower Care
Systemic Rose & Flower Care
Systemic Rose & Flower Food
Worm Ender

BEECH
Imidan
Sevin 5 Dust
Sevin 10 Dust

BEETS
5% Diazinon Granules
Bioneem
Cutworm Earwig & Sowbug Bait
Cygon 2E
Diazinon 2% Granules
Diazinon 5G
Diazinon 5% G
Diazinon 5% Granules
Diazinon 12 1/2% E
Diazinon 25%
Diazinon Dust
Diazinon Plus
Diazinon Soil & Turf Insect Control
Diazinon Spray
Earwig Sowbug & Grasshopper Bait
Fruit & Berry Insect Spray
Grasshopper Earwig & Sowbug Bait
Japanese Beetle Repellent
Liquid Flowable Sevin
Liquid Sevin
Malathion 50%
Malathion 50 Plus
Sevin 5% Dust
Sevin Dura Dust
Sevin Dura-Spray
Sevin Granules
Sevin Liquid
Sevin Liquid/Spray
Sun Spray Ultra Foam
Worm Ender

BEGONIAS
Aphid Mite & Whitefly Killer II
Fruits & Vegetable Insect Spray
Garden Spray
Gentle Care House Plant Spray
Home Patrol Insect Killer
House Plant & Garden Insect Spray
Intercept H&G
Intercept RFO
Intercept Vegetable & Garden Spray

42

INSECTICIDES

USAGE OF PRODUCTS

Japanese Beetle Killer
Mite Beater
Natural Pyrethrin
Permethrin T/O
Plant Insect Control
Rose & Floral Insect Control
Rose & Floral Insect Killer
Roses & Flower Insect Spray
Rose & Flower Insect Killer
Rose & Flower Insect Killer II
Rose, Flower & Ornamental
   Insect Spray
Rose & Garden Insect Fogger
Spider Mite Control
Spider Mite & Mealybug Control
Sun Spray Ultra Fine
Tomato & Vegetable Insect Killer
Whitefly & Mealybug Control
Whitefly & Mealybug Spray
Yard & Garden Insect Killer

BIRCH

2% Systemic Granules
25% Diazinon
Bagworm & Mite Spray
Borer Miner Killer 5
Cygon 2E
Cygon 2-W
Diazinon 4D
Diazinon 25%
Diazinon 25% EC
Diazinon Plus
Diazinon Spray
Dormant Spray & Summer Spray
Horticultural Spray Oil
Imidan
Intercept H&G
Liquid Sevin
Permethrin T/O
Sevin 5 Dust
Sevin 10 Dust
Spray Oil
Systemic Granules
Systemic House Plant
   Insect Control
Systemic Rose & Flower Food

BITTERSWEET

Horticultural Spray Oil

BLACK-EYED SUSAN

Intercept RFO
Whitefly & Mealybug Spray

BLEEDING HEART

Intercept Vegetable & Garden
   Spray

BLUEBERRIES

5% Sevin Garden Dust
25% Methoxychlor Spray
Diazinon Spray
Fruit & Berry Insect Spray
Garden Insect Spray
Methoxychlor 25%
Organic Greenhouse
   House & Veg. Spray
Saf-T-Side
Sevin 5% Dust
Sevin 10 Dust
Sevin Dura Spray
Sevin Granules
Sevin Liquid
Sun Spray Ultra Fine
Thiodan Insect Spray
Worm Ender

BOK CHOY

Bioneem
Bug-B-gon
Diazinon 25%
Diazinon Plus
Flower Garden Insecticidal Soap
Insecticidal Soap Multi Purpose
   Insect Killer
Japanese Beetle Repellent
Worm Ender

BOX ELDER

Sevin 50 WP
Thiodan Insect Spray

BOXWOOD

Bagworm & Mite Spray
Cygon 2E
Diazinon 4E
Diazinon 12 1/2% E
Diazinon 25%
Diazinon 25% EC
Diazinon Plus
Diazinon Spray
Dormant Spray & Summer spray
Intercept H&G
Intercept RFO
Lindane Borer Spray
Sevin 5 Dust
Sevin 10 Dust

BRIDAL VEIL

Intercept RFO
Spider Mite & Meelybug Control
Whitefly & Meelybug Spray

BROCCOLI

5% Diazinon Granules
5% Malathion Dust
5% Sevin Garden Dust
25% Diazinon Insect Spray
50% Malathion Spray
Ant Flea & Tick Insect Granules
Bacillus thuringiensis
Bioneem
Bio Worm Killer
Blue Dragon Garden Dust
Bug-B-Gon
Bug Buster
Cutworm Earwig & Sowbug Bait
Cygon 2E
Diazinon 2% Granules
Diazinon 5G
Diazinon 5% G
Diazinon 5% Granules
Diazinon 25%
Diazinon 25% EC
Diazinon Dust
Diazinon Granules
Diazinon Plus
Diazinon Soil & Turf Insect Control
Dipel
Dipel Dust
Dursban 1% Granules
Dursban Ant & Turf
Dursban Granules
Dursban Lawn & Perimeter
   Granules
Earwig Sowbug & Grasshopper
   Bait
Flea & Tick Granules
Flower Garden Insecticidal Soap
Garden Insect Spray
Garden Soil Insecticide
Garden Spray
Grasshopper Earwig & Sowbug
   Bait
Insecticidal Soap Multi Purpose
   Insect Killer
Intercept Vegetable & Garden
   Spray
Japanese Beetle Killer
Japanese Beetle Repellent
Liquid Flowable Sevin
Malathion 50%
Malathion 50 Plus
Mole Cricket Killer
Multi Purpose Insect Killer
Natural Pyrethrin
Organic Greenhouse
   House & Veg. Spray
Permethrin T/O
Sevin 5 Dust
Sevin 10 Dust
Sevin 5% Dust
Sevin 10% Dust
Sevin 50% WP
Sevin 50 Wettable
Sevin Dura-Dust
Sevin Dura-Spray
Sevin Garden Dust
Sevin Granules
Sevin Liquid
Sevin Spray/Liquid
Systemic Granules
Thiodan
Thiodan Vegetable & Ornamental
   Dust
Thuricide

43

INSECTICIDES

Tomato & Veg. Dust
Tomato & Veg. Fogger
Tomato & Vegetable Insect Killer
Tomato & Veg. RTU
Tomato Pepper Veg. Spray RTU
Vegetable Insect Control
Vegetable Insect Killer
Worm Ender
Yard & Garden Insect Killer

BRUSSELS SPROUTS

5% Diazinon Granules
50% Malathion Spray
Ant Flea & Tick Insect Granules
Bacillus thuringiensis
Bioneem
Blue Dragon Garden Dust
Cutworm Earwig & Sowbug Bait
Diazinon 2% Granules
Diazinon 5G
Diazinon 5% G
Diazinon 5% Granules
Diazinon Soil & Turf Insect Control
Dursban 1% Granules
Dursban Ant & Turf
Dursban Granules
Dursban Lawn & Perimeter Granule
Earwig Sowbug & Grasshopper Bait
Flea & Tick Granules
Flower Garden Insecticidal Soap
Garden Insect Spray
Garden Spray
Grasshopper Earwig & Sowbug Bait
Insecticidal Soap Multi Purpose Insect Killer
Intercept Vegetable & Garden Spray
Japanese Beetle Killer
Japanese Beetle Repellent
Liquid Flowable Sevin
Malathion 50%
Malathion 50 Plus
Natural Pyrethrin
Organic Greenhouse House & Veg. Spray
Permethrin T/O
Sevin 5 Dust
Sevin 10 Dust
Sevin 5% Dust
Sevin 10% Dust
Sevin 50 Wettable
Sevin 50 WP
Sevin Dura-Dust
Sevin Dura-Spray
Sevin Garden Dust
Sevin Granules
Sevin Liquid
Sevin Spray/Liquid
Systemic Granules

Thiodan Insect Spray
Tomato & Veg. Fogger
Tomato & Vegetable Insect Killer
Tomato & Veg. RTU
Tomato Pepper Veg. Spray RTU
Thuricide
Vegetable Insect Control
Vegetable Insect Killer
Worm Ender
Yard & Garden Insect Killer

BUGLEWEED (Aguga)

Intercept RFO

BUILDINGS, PERIMETER TREATMENT

Bug Buster

BULBS (general)

Feed & Shield Systemic Rose Care
Rose & Flower Care
Systemic Rose Care
Systemic Rose & Flower Care
Systemic Rose Shrub & Flower Care
Systemic Rose & Flower Food

CABBAGE

5% Diazinon Granules
5% Malathion Dust
5% Sevin Garden Dust
25% Diazinon
25% Methoxychlor Insect Spray
50% Malathion
Ant Flea & Tick Insect Granules
Bacillus thuringiensis
Bioneem
Bio Worm Killer
Blue Dragon Garden Dust
Bug-B-Gon
Bug Buster
Cutworm Earwig & Sowbug Bait
Cygon 2E
Diazinon 2% Granules
Diazinon 5G
Diazinon 5% G
Diazinon 5% Granules
Diazinon 12 1/2% E
Diazinon 25%
Diazinon 25% EC
Diazinon Dust
Diazinon Granules
Diazinon Plus
Diazinon Soil & Turf Insect Control
Diazinon Spray
Dipel
Dipel Dust
Dursban 1% Granules
Dursban Ant & Turf
Dursban Granules
Dursban Lawn & Perimeter Granule

Earwig Sowbug & Grasshopper Bait
Flea & Tick Granules
Flower Garden Insecticidal Soap
Fruit & Berry Insect Spray
Fruit & Veg Insect Spray
Garden Insect Spray
Garden Soil Insecticide
Garden Spray
Grasshopper Earwig & Sowbug Bait
House Plant & Garden Insect Spray
Insecticidal Soap Multi Purpose Insect Killer
Intercept Vegetable & Garden Spray
Japanese Beetle Killer
Japanese Beetle Repellent
Liquid Flowable Sevin
Malathion 50%
Malathion 50 Plus
Many Purpose Insect Killer
Methoxychlor 25%
Mole Cricket Killer
Multi Purpose Insect Killer
Natural Pyrethrin
Organic Greenhouse House & Veg. Spray
Permethrin T/O
Roses & Flower Insect Spray
Saf-T-Side
Sevin 5% Dust
Sevin 10 Dust
Sevin 10% Dust
Sevin 50 WP
Sevin 50% WP
Sevin 50 Wettable
Sevin Dura-Dust
Sevin Dura-Spray
Sevin Garden Dust
Sevin Granules
Sevin Liquid
Sevin Spray/Liquid
Sun Spray Ultra Fine
Systemic Granules
Thiodan
Thiodan Vegetable & Ornamental Dust
Thuricide
Tomato & Veg. Dust
Tomato & Veg. Fogger
Tomato & Vegetable Insect Killer
Tomato & Veg. RTU
Tomato Pepper Veg. Spray RTU
Vegetable Insect Control
Vegetable Insect Killer
Worm Ender
Yard & Garden Insect Killer

INSECTICIDES

USAGE OF PRODUCTS

CACTUS
Intercept RFO
Whitefly & Mealybug Spray

CALENDULA
2% Systemic Granules
Intercept H&G
Intercept RFO
Spider Mite & Meelybug Control
Systemic Granules
Systemic House Plant
   Insect Control
Whitefly & Meelybug Spray

CAMELLIA
2% Systemic Granules
25% Diazinon
50% Malathion Insect Spray
Bagworm & Mite Spray
Bug-B-Gon
Bug Off Rose & Flower Spray
Cygon 2E
Diazinon 4E
Diazinon 25%
Diazinon 25% EC
Diazinon Plus
Diazinon Spray
Dormant & Summer Oil Spray
Dormant Spray
Dormant Spray & Summer Spray
Feed & Shield Systemic Rose Care
Garden Spray
Gentle Care House Plant Spray
Home Patrol Insect Killer
Horticultural Spray Oil
Intercept H&G
Japanese Beetle Killer
Malathion 50 Plus
Malathion-Oil
Mite Beater
Mite & Insect Spray
Multi Purpose Insect Killer
Natural Pyrethrin
Plant Insect Control
Rose & Floral Insect Control
Rose & Floral Insect Killer
Rose & Flower Insect Killer
Rose & Flower Insect Killer II
Rose, Flower & Ornamental
   Insect Spray
Rose & Flower Spray
Rose & Garden Insect Fogger
Saf-T-Side
Sevin 5 Dust
Sevin 10 Dust
Soluable Oil Spray
Summer & Dormant Oil
Sun Spray Ultra Fine
Systemic Granules
Systemic House Plant
   Insect Control

Systemic Rose & Flower Food
Systemic Shrub & Flower
   Insecticide
Tomato & Vegetable Insect Killer
Vegetable Insect Control
Vegetable Insect Killer
Volck Oil Spray
Yard & Garden Insect Killer

CANEBERRIES
5% Sevin Garden Dust
25% Methoxychlor Spray
Bug-B-Gon
Diazinon 25%
Diazinon 25% EC
Diazinon Plus
Malathion 50%
Methoxychlor 25%
Organic Greenhouse
   House & Veg. Spray
Sevin Granules
Sevin Liquid
Sevin 5% Dust
Worm Ender

CARDON
Worm Ender

CARNATION
2% Systemic Granules
5% Diazinon Dust
5% Sevin Garden Dust
25% Diazinon
50% Malathion
Ant Killer Powder
Bagworm & Mite Spray
Bug-B-Gon
Bug Off Rose & Flower Spray
Cygon 2E
Diazinon 4E
Diazinon 12 1/2% E
Diazinon 25%
Diazinon 25% EC
Diazinon Plus
Diazinon Spray
Garden Spray
Gentle Care House Plant Spray
Home Patrol Insect Killer
Intercept RFO
Intercept Vegetable & Garden
   Spray
Japanese Beetle Killer
Liquid Sevin
Mite Beater
Multi Purpose Insect Killer
Natural Pyrethrin
Plant Insect Control
Rose & Floral Insect Control
Rose & Floral Insect Killer
Rose & Flower Insect Killer
Rose & Flower Insect Killer II
Rose & Flower Spray

Rose, Flower & Ornamental
   Insect Spray
Rose & Garden Insect Fogger
Saf-T-Side
Sevin 5 Dust
Sevin 10 Dust
Sevin 50% WP
Spider Mite & Meelybug Control
Systemic Granules
Systemic House Plant
   Insect Control
Tomato & Vegetable Insect Killer
Vegetable Insect Control
Vegetable Insect Killer
Whitefly & Meelybug Spray
Yard & Garden Insect Killer

CARROTS
5% Diazinon Granules
25% Diazinon Spray
Bioneem
Blue Dragon Garden Dust
Bug-B-Gon
Bug Buster
Cutworm Earwig & Sowbug Bait
Diazinon 2% Granules
Diazinon 5G
Diazinon 5% G
Diazinon 5% Granules
Diazinon 12 1/2% E
Diazinon 25%
Diazinon 25% EC
Diazinon Dust
Diazinon Granules
Diazinon Plus
Diazinon Soil & Turf Insect Control
Diazinon Spray
Earwig Sowbug & Grasshopper
   Bait
Flower Garden Insecticidal Soap
Fruit & Berry Insect Spray
Fruit & Veg Insect Spray
Garden Insect Spray
Garden Spray
Grasshopper Earwig & Sowbug
   Bait
House Plant & Garden
   Insect Spray
Insecticidal Soap Multi Purpose
   Insect Killer
Japanese Beetle Repellent
Liquid Flowable Sevin
Liquid Sevin
Malathion 50%
Many Purpose Insect Killer
Mole Criket Killer
Organic Greenhouse
   House & Veg. Spray
Roses & Flower Insect Spray
Sevin 5% Dust
Sevin 10% Dust
Sevin 10 Dust

45

USAGE OF PRODUCTS

INSECTICIDES

Sevin 50 Wettable
Sevin 50% WP
Sevin 50 WP
Sevin Dura-Dust
Sevin Dura-Spray
Sevin Garden Dust
Sevin Granules
Sevin Liquid
Sevin Spray/Liquid
Thiodan Insect Spray
Thiodan Vegetable & Ornamental Dust
Tomato & Veg. Dust
Worm Ender

CASHEWS

Bioneem
Japanese Beetle Repellent

CATS

5% Sevin Garden Dust
Blue Dragon Garden Dust
Flea-B-Gon Pet Flea & Tick Killer
Sevin 5% Dust
Sevin 5 Dust
Sevin 10 Dust
Sevin Dura Dust
Sevin Garden Dust

CAULIFLOWER

5% Diazinon Granules
5% Malathion Dust
5% Sevin Garden Dust
25% Diazinon Insect Spray
25% Methoxychlor Spray
50% Malathion Spray
Ant Flea & Tick Insect Granules
Bacillus thuringiensis
Bioneem
Bio Worm Killer
Bug-B-Gon
Bug Buster
Cutworm Earwig & Sowbug Bait
Cygon 2E
Diazinon 2% Granules
Diazinon 5G
Diazinon 5% G
Diazinon 5% Granules
Diazinon 25%
Diazinon 25% EC
Diazinon Dust
Diazinon Granules
Diazinon Plus
Diazinon Soil & Turf Insect Control
Dipel
Dipel Dust
Dursban 1% Granules
Dursban Ant & Turf
Dursban Granules
Dursban Lawn & Garden Spray
Earwig Sowbug & Grasshopper Bait
Flea & Tick Granules
Garden Insect Spray
Garden Soil Insecticide
Garden Spray
Grasshopper Earwig & Sowbug Bait
Insecticidal Soap Multi Purpose Insect Killer
Intercept Vegetable & Garden Spray
Japanese Beetle Killer
Japanese Beetle Repellent
Liquid Flowable Sevin
Malathion 50%
Malathion 50 Plus
Many Purpose Insect Killer
Methoxychlor 25%
Mole Cricket Killer
Multi Purpose Insect Killer
Organic Greenhouse House & Veg. Spray
Permethrin T/O
Saf-T-Side
Sevin 5% Dust
Sevin 10% Dust
Sevin 10 Dust
Sevin 50 WP
Sevin Dura-Dust
Sevin Dura-Spray
Sevin Garden Dust
Sevin Granules
Sevin Liquid
Sevin Spray/Liquid
Sun Spray Ultra Fine
Systemic Granules
Thiodan
Thuricide
Tomato & Veg. Dust
Tomato & Veg. Fogger
Tomato & Vegetable Insect Killer
Tomato & Veg. RTU
Tomato Pepper Veg. Spray RTU
Vegetable Insect Control
Vegetable Insect Killer
Worm Ender
Yard & Garden Insect Killer

CEANOTHIS

Sevin 5 Dust
Sevin 10 Dust

CEDAR

Cygon 2E
Horticultural Spray Oil
Imidan
Intercept H&G
Sevin 5 Dust
Sevin 10 Dust
Systemic Shrub & Flower Insecticide

CELERIAC

Worm Ender

CELERY

5% Diazinon Granule
50% Malathion
Bacillus thuringiensis
Bioneem
Bio Worm Killer
Diazinon 2% Granules
Diazinon 5G
Diazinon 5% G
Diazinon 5% Granules
Diazinon Granules
Diazinon Soil & Turf Insect Control
Dipel
Dipel Dust
Flower Garden Insecticidal Soap
Garden Insect Spray
Garden Soil Insecticide
Garden Spray
Insecticidal Soap Multi Purpose Insect Killer
Intercept Vegetable & Garden Spray
Japanese Beetle Killer
Japanese Beetle Repellent
Liquid Flowable Sevin
Malathion 50%
Malathion 50 Plus
Many Purpose Insect Killer
Mole Cricket Killer
Natural Pyrethrin
Organic Greenhouse House & Veg. Spray
Permethrin T/O
Saf-T-Side
Sevin Granules
Sevin Liquid
Sevin Spray/Liquid
Sun Spary Ultra Fine
Thiodan Insect Spray
Thiodan Vegetable & Ornamental Dust
Thuricide
Tomato & Veg. Fogger
Tomato & Vegetable Insect Killer
Tomato & Veg Insect Spray
Tomato & Veg. RTU
Tomato Pepper Veg. Spray RTU
Vegetable Insect Control
Vegetable Insect Killer
Worm Ender

CELOSIA (Cockscomb)

Intercept H&G
Intercept Vegetable & Garden Spray

INSECTICIDES

CEPHALANTHUS
Intercept H&G

CERCIS
Intercept H&G

CHAYOTE
Bug Buster-O

CHERRY
25% Diazinon
25% Methoxychlor Insect Spray
50% Malathion
Bioneem
Borer Miner Killer 5
Bug-B-Gon
Bug Buster-O
Diazinon 12 1/2% E
Diazinon 25%
Diazinon 25% EC
Diazinon Plus
Dormant & Summer Oil Spray
Dormant Oil Spray
Fruit &a Berry Insect Spray
Garden Insect Spray
Horticultural Spray Oil
Imidam
Japanese Beetle Repellent
Lindane Borer Spray
Liquid Flowable Sevin
Malathion 50%
Malathion 50 Plus
Methoxychlor 25%
Organic Greenhouse
   House & Veg. Spray
Permethrin T/O
Saf-T-Side
Scale Away
Sevin 50% WP
Sevin 50 WP
Sevin Dura Spray
Sevin Liquid
Sevin Spray Liquid
Spray Oil
Summer & Dormant Oil
Sun Spray Ultra Fine
Thiodan Insect Spray
Volck Oil Spary
Worm Ender

CHERRY (flowering)
Intercept H&G
Rose & Florel Spray Bomb

CHESTNUT
Worm Ender

CHICKENS
Sevin 5 Dust
Sevin 10 Dust
Sevin Dura Dust

CHICK PEAS
Worm Ender

CHICORY
Worm Ender

CHINABERRY
Dormant Spray & Summer Spray

CHINESE BROCCOLI
25% Diazinon Insect Spray
Bug-B-Gon
Diazinon 25%
Diazinon 25% EC
Diazinon Plus

CHINESE CABBAGE
25% Diazinon Insect Spray
Ant Flea & Tick Insect Granules
Bug-B-Gon
Cutworm Earwig & Sowbug Bait
Diazinon 25% EC
Diazinon 25%
Diazinon Plus
Dursban 1% Granules
Dursban Ant & Turf
Dursban Granules
Dursban Lawn & Perimeter
   Garden
Earwig Sowbug & Grasshopper
   Bait
Flea & Tick Granules
Grasshopper Earwig & Sowbug
   Bait
Liquid Flowable Sevin
Liquid Sevin
Sevin 5% Dust
Sevin Granules
Sevin Liquid
Sevin Spray/Liquid
Worm Ender

CHINESE ELM
Sevin 5 Dust
Sevin 10 Dust

CHIVES
Bioneem
Flower Garden Insecticidal Soap
Insecticidal Soap Multi Purpose
   Insect Killer
Japanese Beetle Repellent

CHRISTMAS TREES (general)
Sun Spray Ultra Fine

CHRYSANTHEMUM
2% Systemic Granules
5% Diazinon Dust
25% Diazinon
50% Malathion
Ant Killer Powder
Bay Worm & Mite Spray
Bug-B-Gon
Diazinon 4E
Diazinon 12 1/2% E
Diazinon 25%
Diazinon 25% EC
Diazinon Plus
Diazinon Spray
Dipel
Fruit & Veg Insect Spray
Garden Spray
Gentle Care House Plant Spray
Home Patrol Insect Killer
House Plant & Garden
   Insect Spray
Intercept H&G
Intercept RFO
Intercept Vegetable & Garden
   Spray
Japanese Beetle Killer
Lindane Borer Spray
Liquid Sevin
Malathion 50 Plus
Mite Beater
Multi Purpose Insect Killer
Natural Pyrethrin
Plant Insect Control
Rose & Floral Insect Control
Rose & Floral Insect Killer
Rose & Floral Spray Boost
Roses & Flower Insect Spray
Rose & Flower Insect Killer
Rose & Flower Insect Killer II
Rose, Flower & Ornamental
   Insect Spray
Rose & Garden Insect Fogger
Sevin 5 Dust
Sevin 10 Dust
Sun Spray Ultra Fine
Systemic Granules
Saf-T-Side
Sevin 50% WP
Systemic House Plant
   Insect Control
Tomato & Vegetable Insect Killer
Vegetable Insect Control
Vegetable Insect Killer
Whitefly & Mealybug Spray
Yard & Garden Insect Killer

CITRUS (general)
5% Malathion Dust
5% Sevin Garden Dust
50% Malathion
Bacillus thuringiensis
Bioneem
Bio Worm Killer
Bug Buster-O
Diazinon Spray
Diazinon 12 1/2% E
Dormant Spray

USAGE OF PRODUCTS

INSECTICIDES

Dormant Spray & Summer Spray
Dormant & Summer Oil Spray
Flower Garden Insecticidal Soap
Fruit & Berry Insect Spray
Fruit & Veg Insect Spray
Horticultural Spray Oil
House Plant & Garden
 Insect Spray
Insecticidal Soap Multi Purpose
 Insect Killer
Japanese Beetle Repellent
Kelthane
Liquid Flowable Sevin
Malathion 50%
Malathion 50 Plus
Natural Pyrethrin
Organic Greenhouse
 House & Veg Spray
Permethrin T/O
Red Spider Spray
Roses & Flower Insect Spray
Saf-T-Side
Scale Away
Sevin 50% WP
Sevin Dura Spray
Sevin Liquid
Sevin Spray/Liquid
Soluable Oil Spray
Spray Oil
Summer Dormant Oil
Sun Spray Ultra Fine
Thuricide
Volck Oil Spray
Worm Ender

COFFEE

Bug Buster-O

COLE CROPS (general)

Bug Buster-O
Saf-T-Side
Sun Spray Ultra Fine

COLEUS

Aphid Mite & Whitefly Killer II
Bug-B-Gon
Intercept H&G
Intercept RFO
Intercept Vegetable & Garden
 Spray
Multi Purpose Insect Killer
Permethrin T/O
Spider Mite Control
Spider Mite & Meelybug Control
Whitefly & Meelybug Control
Whitefly & Meelybug Spray

COLLARDS

5% Diazinon Granules
5% Malathion Dust
25% Diazinon Insect Spray
Ant, Flea & Tick Insect Granules
Bacillus thuringiensis
Bioneem
Bio Worm Killer
Blue Dragon Garden Dust
Bug-B-Gon
Bug Buster
Cutworm Earwig & Sowbug Bait
Cygoon 2E
Diazinon 2% Granules
Diazinon 5G
Diazinon 5% G
Diazinon 5% Granules
Diazinon 25%
Diazinon 25% EC
Diazinon Granules
Diazinon Plus
Diazinon Soil & Turf Insect Control
Dipel
Dipel Dust
Dursban 1% Granules
Dursban Ant & Turf
Dursban Lawn & Perimeter
 Granule
Earwig Sowbug & Grasshopper
 Bait
Flea & Tick Granules
Garden Insect Spray
Garden Soil Insecticide
Garden Spray
Grasshopper Earwig & Sowbug
 Bait
Japanese Beetle Killer
Japanese Beetle Repellent
Liquid Flowable Sevin
Liquid Sevin
Malathion 50 Plus
Mole Cricket Killer
Natural Pyrethrin
Sevin 5% Dust
Sevin Granules
Sevin Liquid
Sevin 50 Wettable
Sevin Spray/Liquid
Thiodam Insect Spray
Thiodan Vegetable & Ornamental
 Dust
Thuricide
Tomato & Veg. Fogger
Tomato & Vegetable Insect Killer
Tomato & Vegetable Insect Spray
Tomato & Veg. RTU
Tomato Pepper Veg. Spray RTU
Vegetable Insect Control
Vegetable Insect Killer
Worm Ender

CONIFERS (general)

Horticultural Spray Oil
Imidan
Permethrin T/O
Saf-T-Side
Sun Spray Ultra Fine

CORALBERRY

Aphid Mite & Whitefly Killer II
Spider Mite Control
Whitefly & Mealybug Control

CORN (Sweet)

5% Diazinon Granules
25% Diazinon Spray
Ant Flea & Tick Insect Granules
Bioneem
Blue Dragon Garden Dust
Bug Buster
Cutworm Earwig & Sowbug Bait
Diazinon 2% Granules
Diazinon 5% G
Diazinon 5G
Diazinon 5% Granules
Diazinon 12 1/2% E
Diazinon 25%
Diazinon 25% EC
Diazinon Dust
Diazinon Granules
Diazinon Plus
Diazinon Soil & Turf Insect Control
Diazinon Spray
Dursban 1% Granules
Earwig Sowbug & Grasshopper
 Bait
Flea & Tick Granules
Fruit & Berry Insect Spray
Fruit & Veg Insect Spray
Garden Insect Spray
Garden Soil Insecticide
Grasshopper Earwig & Sowbug
 Bait
House Plant & Garden
 Insect Spray
Intercept Vegetable & Garden
 Spray
Japanese Beetle Repellent
Liquid Flowable Sevin
Malathion 50%
Mole Cricket Killer
Multi Purpose Insect Killer
Permethrin T/O
Roses & Flower Insect Spray
Saf-T-Side
Sevin 5 Dust
Sevin 5% Dust
Sevin 5 Dust
Sevin 10% Dust
Sevin 10 Dust
Sevin 50 WP
Sevin Dura-Dust
Sevin Dura-Spray
Sevin Garden Dust
Sevin Granules
Sevin Liquid
Sevin Spray/Liquid
Sevin Wettable
Sun Spray Ultra Fine

48

INSECTICIDES

Thiodan Vegetable & Ornamental Dust
Tomato & Veg. Dust
Tomato & Veg. Fogger

COTONEASTER

Dormant Oil Spray
Intercept H&G

COWPEAS (Black-eye peas)

5% Malathion Dust
Garden Insect Spray
Liquid Sevin
Malathion 50%
Sevin Liquid
Sevin Spray/Liquid

CRABAPPLES

Aphid Mite & Whitefly Killer
Bug Buster-O
Horticultural Spray Spray
Intercept H&G
Kelthane
Rose & Florel Spray Bomb
Rose & Flower Insect Killer II

CRANBERRIES

Diazinon Spray
Fruit & Berry Insect Spray
Fruit & Veg Insect Spray
Garden Spray
House Plant & Garden Insect Spray
Japanese Beetle Killer
Methoxychlor 25%
Natural Pyrethrin
Roses & Flower Insect Spray
Sevin Dura Spray
Sevin Liquid
Tomato & Veg. Fogger
Tomato & Veg Insect Spray
Tomato & Veg RTU
Tomato Pepper Veg. Spray RTU

CRAPE MYRTLE

Intercept H&G
Rose & Florel Spray Bomb

CROTONS

Soluable Oil Spray

CROWN OF THORNS

Intercept Vegetable & Garden Spray
Sun Spray Ultra Fine

CUCUMBERS

5% Diazinon Granule
5% Malathion Dust
5% Sevin Garden Dust
25% Diazinon
50% Malathion

Bacillus thuringiensis
Bioneem
Blue Dragon Garden Dust
Bug-B-Gon
Bug Buster
Cutworm Earwig & Sowbug Bait
Diazinon 2% Granules
Diazinon 5G
Diazinon 5% G
Diazinon 5% Granules
Diazinon 12 1/2% E
Diazinon 25%
Diazinon 25% EC
Diazinon Dust
Diazinon Granules
Diazinon Plus
Diazinon Soil & Turf Insect Control
Diazinon Spray
Dipel
Dipel Dust
Earwig Sowbug & Grasshopper Bait
Flower Garden Insecticidal Soap
Fruit & Berry Insect Spray
Fruit & Veg Insect Spray
Garden Insect Spray
Garden Soil Insecticide
Grasshopper Earwig & Sowbug Bait
House Plant & Garden Insect Spray
Insecticidal Soap Multi Purpose Insect Killer
Japanese Beetle Repellent
Kelthane
Liquid Flowable Sevin
Liquid Sevin
Malathion 50%
Malathion 50 Plus
Many Purpose Insect Killer
Mole Cricket Killer
Multi Purpose Insect Killer
Roses & Flower Insect Spray
Saf-T-Side
Sevin 5 Dust
Sevin 10 Dust
Sevin 5% Dust
Sevin 10% Dust
Sevin Garden Dust
Sevin Granules
Sevin Liquid
Sevin Wettable
Sevin 50% WP
Sevin 50 WP
Sevin Dura-Dust
Sevin Dura-Spray
Sevin Spray/Liquid
Thiodan Insect Spray
Thiodan Vegetable & Ornamental Dust
Thuricide
Tomato & Vegetable Insect Killer

Worm Ender
Yard & Garden Insect Killer

CUCURBITS (general)

Sun Spray Ultra Fine
Worm Ender

CURRANTS

Worm Ender

CYCLAMEN

Intercept Vegetable & Garden Spray

CYPRESS

Cygon 2E
Intercept Vegetable & Garden Spray
Sevin 5 Dust
Sevin 10 Dust
Systemic Shrub & Flower Insecticide

DAFFODIL

Intercept H&G

DAHLIAS

5% Diazinon Dust
Ant Killer Powder
Bug Off Rose & Flower Spray
Garden Spray
Gentle Care House Plant Spray
Home Patrol Insect Killer
Intercept H&G
Japanese Beetle Killer
Mite Beater
Natural Pyrethrin
Plant Insect Control
Rose & Floral Insect Control
Rose & Floral Insect Killer
Rose & Flower Insect Killer
Rose & Flower Spray
Rose, Flower & Ornamental Insect Spray
Rose & Garden Insect Fogger
Sevin 5 Dust
Sevin 10 Dust
Systemic Granules
Systemic House Plant Insect Control
Tomato & Vegetable Insect Killer
Vegetable Insect Control
Vegetable Insect Killer

DAISY

Intercept RFO
Whitefly & Mealybug Spray

DAISY-SHASTA

Aphid Mite & Whitefly Killer II
Intercept H&G

USAGE OF PRODUCTS

49

INSECTICIDES

Spider Mite Control
Whitefly & Mealybug Control

DANDELIONS

Liquid Flowable Sevin
Liquid Sevin
Sevin 5% Dust
Sevin Granules
Sevin Liquid
SevinSpray/Liquid

DAPHNE

Intercept RFO
Spider Mite & Meelybug Control
Whitefly & Meelybug Spray

DAYLILIES

Cygon 2E
Fruit & Veg Insect Spray
House Plant & Garden
   Insect Spray
Roses & Flower Insect Spray
Systemic Shrub & Flower
   Insecticide

DELPHINIUM

2% Systemic Granules
Mite Beater
Sevin 5 Dust
Sevin 10 Dust
Systemic Granules
Systemic House Plant
   Insect Control

DEWBERRIES

Diazinon 25%

DIANTHUS

Intercept H&G

DICHONDRA

Dursban Insect Spray

DIEFFENBACHIA (Dumbcane)

Aphid Mite & Whitefly Killer II
Fruit & Veg Insect Spray
House Plant & Garden
   Insect Spray
Intercept H&G
Intercept Vegetable & Garden
   Spray
Roses & Flower Insect Spray
Spider Mite Control
Spider Mite & Mealybug Control
Sun Spray Ultra Fine
Whitefly & Mealybug Control

DILL

Bioneem
Flower Garden Insecticidal Soap
Insecticidal Soap Multi Purpose
Insect Killer
Japanese Beetle Repellent

DOGS

5% Malathion Dust
5% Sevin Garden Dust
50% Malathion
Blue Dragon Garden Dust
Flea-B-Gon Pet Flea & Tick Killer
Flea, Tick & Mange Dip
Sevin 5 Dust
Sevin 10 Dust
Sevin 5% Dust
Sevin Dura Dust
Sevin Garden Dust

DOG HOUSES

Flea & Tick Spray
Flying Insect Killer

DOGWOOD

25% Diazinon Insect Spray
Borer Miner Killer 5
Dormant Spray & Summer Spray
Fruits & Vegetable Insect Spray
Garden Insect Spray
Garden Spray
Gentle Care House Plant Spray
Home Patrol Insect Killer
Horticultural Spray Oil
House Plant & Garden
   Insect Spray
Imidan
Intercept RFO
Japanese Beetle Killer
Lindane Borer Spray
Mite Beater
Natural Pyrethrin
Liquid Sevin
Plant Insect Control
Rose & Floral Insect Control
Rose & Floral Insect Killer
Roses & Flower Insect Spray
Rose & Flower Insect Killer
Rose & Flower Insect Killer II
Rose, Flower & Ornamental
   Insect Spray
Rose & Garden Insect Fogger
Sevin 5 Dust
Sevin 10 Dust
Spray Oil
Thiodan Insect Spray
Tomato & Vegetable Insect Killer
Vegetable Insect Control
Vegetable Insect Killer

DOUGLAS FIR

25% Diazinon Insect Spray
Bagworm & Mite Spray
Diazinon 25%
Diazinon 25% EC
Diazinon Spray

DRACAENA

Aphid Mite & Whitefly Killer II
Fruit & Veg Insect Spray
House Plant & Garden
   Insect Spray
Intercept RFO
Intercept Vegetable & Garden
   Spray
Roses & Flower Insect Spray
Spider Mite Control
Spider Mite & Mealybug Control
Sun Spray Ultra Fine
Systemic Rose & Flower Food
Whitefly & Mealybug Control
Whitefly & Mealybug Spray

EASTER LILY

Intercept RFO
Systemic House Plant
   Insect Control
Sun Spray Ultra Fine
Systemic Granules
Whitefly & Mealybug Spray

EGGPLANT

5% Sevin Garden Dust
25% Methoxychlor Spray
50% Malathion
Bioneem
Blue Dragon Garden Dust
Bug Buster
Colo Potato Beetle Beater
Cutworm Earwig & Sowbug Bait
Earwig Sowbug & Grasshopper
   Bait
Flower Garden Insecticidal Soap
Garden Insect Spray
Garden Spray
Grasshopper Earwig & Sowbug
   Bait
Insecticidal Soap Multi Purpose
   Insect Killer
Intercept Vegetable & Garden
   Spray
Japanese Beetle Killer
Japanese Beetle Repellent
Liquid Flowable Sevin
Liquid Sevin
Malathion 50%
Malathion 50 Plus
Methoxychlor 25%
Natural Pyrethrin
Organic Greenhouse
   House & Veg. Spray
Permethrin T/O
Saf-T-Side
Sevin 5 Dust
Sevin 10 Dust
Sevin 5% Dust
Sevin 10% Dust
Sevin 50 WP

INSECTICIDES

Sevin 50 Wettable
Sevin Dura Dust
Sevin Dura Spray
Sevin Garden Dust
Sevin Granules
Sevin Liquid
Sevin Spray/Liquid
Sun Spray Ultra Fine
Thiodan Insect Spray
Thiodan Vegetable & Ornamental Dust
Tomato & Veg. Dust
Tomato & Vegetable Insect Killer
Tomato & Veg RTU
Tomato Pepper Veg. Spray RTU
Vegetable Insect Control
Vegetable Insect Killer
Worm Ender
Yard & Garden Insect Killer

ELM

25% Diazinon Insect Spray
Bagworm & Mite Spray
Borer Killer II
Borer Miner Killer 5
Colo. Potato Beetle Beater
Diazinon 4E
Diazinon 25%
Diazinon 25% EC
Diazinon Spray
Dormant Oil Spray
Horticultural Spray Oil
Imidan
Intercept RFO
Intercept Vegetable & Garden Spray
Liquid Sevin
Permethrin T/O
Sevin 5 Dust
Sevin 10 Dust
Sevin 50 WP

ENDIVE (escarole)

5% Diazinon Granules
25% Diazinon Insect Spray
Bug-B-Gon
Cutworm Earwig & Sowbug Bait
Cygon 2E
Diazinon 2% Granules
Diazinon 5% G
Diazinon 5% Granules
Diazinon 5G
Diazinon 25%
Diazinon 25% EC
Diazinon Dust
Diazinon Granules
Diazinon Plus
Diazinon Soil & Turf Insect Control
Earwig Sowbug & Grasshopper Bait
Garden Soil Insecticide
Grasshopper Earwig & Sowbug Bait
Liquid Flowable Sevin
Liquid Sevin
Malathion 50 Plus
Mole Cricket Killer
Sevin 5% Dust
Sevin Dura-Spray
Sevin Granules
Sevin Liquid
Sevin Spray/Liquid
Worm Ender

EUONYMUS

2% Systemic Granules
Aphid Mite & Whitefly Killer
Cygon 2E
Dormant Spray & Summer Spray
Home Patrol Insect Killer
Intercept H&G
Intercept Vegetable & Garden Spray
Mite Beater
Permethrin T/O
Rose & Flower Insect Killer II
Sevin 5 Dust
Sevin 10 Dust
Systemic Granules
Systemic Shrub & Flower Insecticide

EVERGREENS (general)

50% Malathion
Dormant Oil Spray
Dormant Spray
Dormant & Summer Oil Spray
Dursban Insecticide
Dursban Lawn Insect Killer
Horticultural Spray Oil
Imidan
Malathion 50%
Malathion 50 Plus
Many Purpose Dursban Concentrate
Mite & Insect Spray
Sevin 50 WP
Sevin 50% WP
Spray Oil
Volck Oil Spray

EXACUM

Intercept Vegetable & Garden Spray
Permethrin T/O

FATSHEDERA

Intercept RFO
Spider Mite & Meelybug Control
Whitefly & Meelybug Spray

FENCE POSTS

Ortho-Klor
Termi-chlor
Termite Killer

FERNS (general)

Aphid Mite & Whitefly Killer
Aphid Mite & Whitefly Killer II
Fruit & Veg Insect Spray
House Plant & Garden Insect Spray
Intercept H&G
Intercept RFO
Roses & Flower Insect Spray
Spider Mite Control
Spider Mite & Mealybug Control
Sun Spray Ultra Fine
Systemic Granules
Whitefly & Mealybug Control
Whitefly & Mealybug Spray

FERN (Birds Nest)

Intercept Vegetable & Garden Spray

FERN (Maidenhair)

Sevin 5 Dust
Sevin 10 Dust

FERN (Rabbits foot)

Intercept Vegetable & Garden Spray

FICUS

Cygon 2E
Intercept H&G
Intercept RFO
Spider Mite & Meelybug Control
Systemic Rose & Flower Food
Systemic Shrub & Flower Insecticide
Whitefly & Meelybug Spray

FIGS

Bioneem
Horticultural Spray Oil
Japanese Beetle Repellent
Spray Oil

FILBERTS

Malathion 50%
Sevin Dura Spray
Sevin Liquid
Sevin Spray/Liquid
Summer & Dormant Oil
Thiodan Vegetable & Ornamental Dust
Worm Ender

USAGE OF PRODUCTS

51

INSECTICIDES

FIR
Diazinon 4E
Diazinon 25%
Imidan
Intercept H&G
Permethrin T/O
Sevin 5 Dust
Sevin 10 Dust

FLOWERS (general)
2% Systemic Granules
25% Diazinon
25% Methoxychlor Insect Spray
50% Malathion
Bagworm & Mite Spray
Bioneem
Bug-B-Gon
Bug Buster-O
Cutworm Earwig & Sowbug Bait
Diazinon 12 1/2% E
Diazinon 25% EC
Diazinon Dust
Dipel
Dursban
Dursban Granules
Dursban Insecticide Con.
Dursban Insect Spray
Dursban Lawn Insect Killer
Earwig Sowbug & Grasshopper Bait
Feed & Shield Systemic Rose Care
Flea & Tick Granules
Garden Spray
House Plant Insecticidal Soap
Insecticidal Soap Multi Purpose Insect Killer
Isotox
Japanese Beetle Killer
Japanese Beetle Repellent
Kelthane
Liquid Flowable Sevion
Liquid Sevin
Malathion 50%
Many Purpose Dursban Concentrate
Methoxychlor 25%
Mite & Insect Spray
Multi Purpose Insect Killer
Orthene
Pestkil
Rose & Florel Spray Bomb
Rose & Flower Care
Rose & Flower Insect Killer
Rose & Flower Insect Killer II
Rose & Flower Insect Spray
Rose & Flower Dust
Rose, Flower & Ornamental Insect Spray
Saf-T-Side
Sevin Dura-Spray
Sevin Granules
Sevin Liquid
Sevin Spray/Liquid
Systemic Granules
Systemic Rose Care
Systemic Rose & Flower Care
Systemic Rose & Flower Food
Systemic Rose Shrub & Flower Care
Worm Ender

FLOWERING PLANTS (general)
Sun Spray Ultra Fine
Systemic Rose & Flower Care

FLUFFY RUFFLES
Aphid Mite & Whitefly Killer II
Spider Mite Control
Whitefly & Mealybug Control

FLY BAIT
Malathion 50%

FOLIAGE PLANTS (general)
Flower Garden Insecticidal Soap
Houseplant Insecticidal Soap
Insecticidal Soap Multi Purpose Insect Killer
Saf-T-Side
Sun Spray Ultra Fine

FORSYTHIA
Intercept RFO

FOUNTAIN GRASS
Intercept H&G

FRUIT TREES (general)
Dormant Spray
Flower Garden Insecticidal Soap
Fruit & Vegetable Insect Killer
Houseplant Insecticidal Soap
Insecticidal Soap Multi Purpose Insect Killer
Insecticidal Soap for Fruit & Vegetables

FUSCHIA
Fruit & Veg Insect Spray
Home Patrol Insect Killer
House Plant & Garden Insect Spray
Intercept H&G
Intercept RFO
Intercept Vegetable & Garden Spray
Rose & Flower Insect Spray
Saf-T-Side
Sevin 5 Dust
Sevin 10 Dust
Spider Mite & Mealybug Control
Whitefly & Mealybug Spray

GARDENIAS
5% Malathion Dust
Cygon 2E
Dormant & Summer Oil Spray
Intercept H&G
Intercept RFO
Malathion-Oil
Rose & Florel Spray Bomb
Soluable Oil Spray
Spider Mite & Mealybug Control
Sun Spray Ultra Fine
Systemic Rose & Flower Food
Systemic Shrub & Flower Insecticide
Volck Oil Spray
Whitefly & Mealybug Spray

GARLIC
Bioneem
Japanese Beetle Repellent
Malathion 50 Plus
Worm Ender

GANZANIA
Intercept H&G

GERANIUMS
Aphid Mite & Whitefly Killer II
Fruits & Vegetable Insect Spray
Garden Spray
Gentle Care House Plant Spray
Home Patrol Insect Killer
House Plant & Garden Insect Spray
Intercept H&G
Intercept RFO
Japanese Beetle Killer
Mite Beater
Natural Pyrethrin
Plant Insect Control
Rose & Floral Insect Control
Rose & Floral Insect Killer
Roses & Flower Insect Spray
Rose & Flower Insect Killer II
Rose, Flower & Ornamental Insect Spray
Rose & Garden Insect Fogger
Spider Mite Control
Spider Mite & Meelybug Control
Sun Spray Ultra Fine
Systemic Rose & Flower Food
Vegetable Insect Control
Vegetable Insect Killer
Whitefly & Meelybug Control
Whitefly & Meelybug Spray

GERBERAS
Cygon 2E
Systemic Shrub & Flower Insecticide

52

INSECTICIDES

USAGE OF PRODUCTS

GINGER
- Bioneem
- Japanese Beetle Repellent

GLADIOLUS
- 5% Sevin Garden Dust
- 25% Diazinon
- Bagworm & Mite Spray
- Bug-B-Gon
- Cygon 2E
- Diazinon 4E
- Diazinon 12 1/2% E
- Diazinon 25%
- Diazinon 25% EC
- Diazinon Plus
- Diazinon Spray
- Garden Spray
- Gentle Care House Plant Spray
- Intercept RFO
- Intercept Vegetable & Garden Spray
- Japanese Beetle Killer
- Liquid Sevin
- Multi Purpose Insect Killer
- Natural Pyrethrin
- Permethrin T/O
- Plant Insect Control
- Rose & Floral Insect Control
- Rose & Floral Insect Killer
- Rose & Flower Insect Killer
- Rose & Flower Insect Killer II
- Rose Flower & Ornamental Insect Spray
- Rose, Flower & Ornamental Insect Spray
- Rose & Garden Insect Fogger
- Saf-T-Side
- Sevin 50% WP
- Systemic Rose & Flower Food
- Systemic Shrub & Flower Insecticide
- Tomato & Vegetable Insect Killer
- Vegetable Insect Control
- Vegetable insect Killer

GLOWERING TREE
- Horticultural Spray Oil

GOLD BELLS
- Intercept Vegetable & Garden Spray
- Permethrin t/O

GRAPES
- 25% Diazinon
- 25% Methoxychlor Spray
- 50% Malathion Spray
- Bacillus thuringiensis
- Bio Worm Killer
- Blue Dragon Garden Dust
- Bug-B-Gon
- Diazinon 12 1/2% E
- Diazinon 25%
- Diazinon 25% EC
- Diazinon Spray
- Dipel
- Dipel Dust
- Dormant Oil Spray
- Fruit & Berry Insect Spray
- Fruit & Veg Insect Spray
- Garden Insect Spray
- Home Patrol Insect Killer
- Horticultural Spray Oil
- House Plant & Garden Insect Spray
- Imidan
- Kelthane
- Liquid Flowable Sevin
- Malathion 50%
- Malathion 50 Plus
- Methoxychlor 25%
- Mite Beater
- Red Spider Spray
- Roses & Flower Insect Spray
- Saf-T-Side
- Sevin 5 Dust
- Sevin 10 Dust
- Sevin 5% Dust
- Sevin 50 Wettable
- Sevin Liquid
- Sevin 50 WP
- Sevin Granules
- Sun Spray Ultra Fine
- Thiodan Vegetable & Ornamental Dust
- Thuricide
- Worm Ender

GRAPE IVY
- Intercept Vegetable & Garden Spray

GREENHOUSES (home)
- Flower Garden Insecticidal Soap
- Houseplant Insecticidal Soap
- Insecticidal Soap Multi Purpose Insect Killer

GREENS (general)
- Worm Ender

GROUND COVERS
- Dursban Insecticide
- Fruit & Veg Insect Spray
- House Plant & Garden Insect Spray
- Roses & Flower Insect Spray

HACKBERRY
- Dormant Spray & Summer Spray

HANOVER SALAD
- Liquid Flowable Sevin
- Sevin Granules
- Sevin Liquid
- Sevin Spray/Liquid

HAWTHORN (Crataegus)
- 25% Diazinon
- Bagworm & Mite Spray
- Diazinon 4E
- Diazinon 25%
- Diazinon 25% EC
- Diazinon Plus
- Diazinon Spray
- Dormant Oil Spray
- Horticultural Spray Oil
- Imidan
- Intercept H&G
- Spray Oil
- Systemic Rose & Flower Food

HELIOTROPE
- Intercept RFO
- Spider Mite & Meelybug Control

HEMLOCK
- Borer Miner Killer 5
- Cygon 2E
- Horticultural Spray Oil
- Imidan
- Intercept H&G
- Intercept RFO
- Sevin 5 Dust
- Sevin 10 Dust
- Systemic Shrub & Flower Insecticide

HERBS (general)
- Fruits & Vegetable Insect Spray
- Houseplant & Garden Insect Spray
- Roses & Flower Insect Spray
- Tomato & Vegetable Insect Killer
- Worm Ender
- Yard & Garden Insect Killer

HERBS & SPICES (general)
- Bug Buster-O

HIBISCUS
- Intercept RFO
- Saf-T-Side
- Soluable Oil Spray
- Spider Mite & Mealybug Control
- Sun Spray Ultra Fine
- Whitefly & Mealybug Spray

HICKORY
- Imidan
- Sevin 5 Dust
- Sevin 10 Dust

53

INSECTICIDES

HOLLY
5% Diazinon Dust
25% Diazinon
Ant Killer Powder
Aphid Mite & Whitefly Killer II
Bagworm & MiteSpray
Cygon 2E
Diazinon 4E
Diazinon12 1/2% E
Diazinon 25%
Diazinon Plus
Diazinon Spray
Dormant Spray & Summer Spray
Feed & Shield Systemic Rose Care
Horticultural Spray Oil
Intercept H&G
Intercept RFO
Sevin 5 Dust
Sevin 10 Dust
Soluable Oil Spray
Spider Mite Control
Spray Oil
Systemic Shrub & Flower Insecticide
Whitefly & Mealybug Control

HOME FOUNDATIONS
Diazinon 5G
Diazinon 25% EC
Diazinon Granules

HONEY LOCUST
Horticultural Spray Oil

HONEYSUCKLE
Isotox
Orthene
Permethrin T/O
Systemic Rose & Flower Food

HORSERADISH
Cutworm Earwig & Sowbug Bait
Earwig Sowbug & Grasshopper Bait
Grasshopper Earwig & Sowbug Bait
Intercept Vegetable & Garden Spray
Liquid Flowable Sevin
Liquid Sevin
Permethrin T/O
Sevin 5% Dust
Sevin Liquid
Sevin Spray/Liquid
Worm Ender

HORNBEAN
Sevin 5 Dust
Sevin 10 Dust

HOUSEPLANTS (general)
2% Systemic Granules
Flower Garden Insecticidal Soap
Houseplant Insecticidal Soap
Insecticidal Soap Multi Purpose Insect Killer

HOYA
Intercept RFO
Spider Mite & Meelybug Control
Whitefly & Meelybug Spray

HYDRANGEA
Intercept H&G
Liquid Sevin
Sevin 5 Dust
Sevin 10 Dust

HYPOESTES
Intercept H&G
Intercept RFO
Intercept Vegetable & Garden Spray
Permethrin T/O

ICE PLANT
Intercept RFO
Spider Mite & Meelybug Control
Whitefly & Meelybug Spray

ILEX
Horticultural Spray Oil
Systemic Rose & Flower Food

IMPATIENS
Intercept H&G
Sun Spray Ultra Fine

INCH PLANT
Intercept RFO
Spider Mite & Meelybug Control
Whitefly & Meelybug Spray

INDOORS (general)
50% Malathion
Alter
Ant Barrier
Ant Dust
Ant Flea & Spider Killer
Ant Killer Dust II
Ant Killer Plus
Ant Roach & Spider Control
Ant Roach & Spider Killer II
Ant-Stop
Ant-Stop RTU
Ant-Stop Bait 2
Ant-Stop Dust
Aphid Mite & Whitefly Killer
Carpenter Ant Control
Carpenter Ant Killer
Diazinon 4E
Diazinon 5% G
Diazinon 12 1/2% E
Diazinon 25%
Dursban 1% Granules
Dursban Insecticide Conc.
Dursban Lawn Insect Killer
Dursban Lawn Insect Spray
Fire Ant Killer
Flea & Tick Granules
Flea Beater
Flea-B-Gon Total
Flea-B-Gon Total Indoor Fogger
Flea-B-Gon Total RTU
Flea-B-Gon Pet Flea & Tick Killer
Flea-B-Gon Flea & Tick Killer
Flea Free Carpet Treatment
Flying Insect Killer
Garden Spray
Home Defense Indoor Outdoor Insect Killer
Home Defense home & Garden Insect Killer
Home Defense Hi-Power
Home Defense Hi Power Roach Ant & Spider Killer
Home Defense Flying & Crawling Insect Killer
Home Pest Control
Home Pest Control Conc.
Home Pest & Carpet Dust
Home Pest Insect Control
Hornet & Wasp Killer II
House Plant Spray II
Household Insect Control
Indoor Flea Killer
Indoor Flea & Tick Spray
Intercept RFO
Liquid Flowable Sevin
Many Purpose Dursban Concentrate
Many Purpose Insect Killer
P.C. Pest Control Conc.
Predator Carpenter Ant Killer Dust
Predator Home Insect Killer II
Predator Roach Powder
Predator Termite Killer Dust
Roach Powder
Roach-Rid Home Pest Killer
Sevin 5% Dust
Sevin 5 Dust
Sevin 10 Dust
Sevin Granules
Striker Roach & Ant Bait
Systemic Rose Care
Systemic Shrub & Flower Insecticide
Termite & Carpenter Ant Control
Termite Killer
Ultra S.S.C.
Vikor Concentrate
Vikor Dual Action
Vikor Ready to Use
Wasp & Hornet Spray

INSECTICIDES

USAGE OF PRODUCTS

INKBERRY
Sevin 5 Dust
Sevin 10 Dust

IRIS
Cygon 2E
Fruit & Veg Insect Spray
House Plant & Garden
  Insect Spray
Intercept H&G
Intercept RFO
Roses & Flower Insect Spray
Saf-T-Side
Sevin 5 Dust
Sevin 10 Dust
Systemic Rose & Flower Food
Thiodan Vegetable & Ornamental
  Dust

IVY
Aphid Mite & Whitefly Killer II
Fruit & Veg Insect Spray
Home Patrol Insect Killer
House Plant & Garden
  Insect Spray
Intercept H&G
Intercept RFO
Intercept Vegetable & Garden Spray
Permethrin T/O
Roses & Flower Insect Spray
Soluable Oil Spray
Spider Mite Control
Spider Mite & Mealybug Control
Whitefly & Mealybug Control
Whitefly & Mealybug Spray

IVY (Boston)
Isotox
Orthene

IVY (English)
Mite Beater
Permethrin T/O

IXORA
Intercept H&G
Intercept RFO
Malathion-Oil
Soluable Oil Spray

JADE PLANT (crassula)
Aphid Mite & Whitefly Killer II
Fruit & Veg Insect Spray
Home Patrol Insect Killer
House Plant & Garden
  Insect Spray
Mite Beater
Roses & Flower Insect Spray
Spider Mite Control
Sun Spray Ultra Fine
Whitefly & Mealybug Control

JOJOBA
Bug Buster-O

JUNIPER
25% Diazinon Insect Spray
Bagworm & Mite Spray
Cygon 2E
Diazinon 4E
Diazinon 25%
Diazinon 25% EC
Diazinon Spray
Dormant Spray & Summer Spray
Horticultural Spray Oil
Imidan
Intercept H&G
Intercept Vegetable & Garden
  Spray
Liquid Sevin
Sevin 5 Dust
Sevin 10 Dust
Systemic Shrub & Flower
  Insecticide

KALANCHAE
Aphid Mite & Whitefly Killer II
Spider Mite Control
Whitefly & Mealybug Control

KALANDRA
Intercept H&G

KALE
5% Diazinon Granules
5% Malathion Dust
25% Diazinon Spray
50% Malathion Spray
Ant Flea & Tick Insect Granules
Bacillus thuingiensis
Bioneem
Bio Worm Killer
Blue Dragon Garden Dust
Bug-B-Gon
Cygon 2E
Diazinon 2% Granules
Diazinon 5G
Diazinon 5% G
Diazinon 5% Granules
Diazinon 25%
Diazinon 25% EC
Diazinon Plus
Diazinon Soil & Turf Insect Control
Dipel
Dipel Dust
Dursban 1% Granules
Dursban Ant & Turf
Dursban Lawn & Perimeter
  Granule
Flea & Tick Granules
Flower Garden Insecticidal Soap
Garden Insect Spray
Garden Soil Insecticide
Garden Spray
Insecticidal Soap Multi Purpose
  Insect Killer
Japanese Beetle Killer
Japanese Beetle Repellent
Liquid Flowable Sevin
Liquid Sevin
Malathion 50%
Malathion 50 Plus
Mole Cricket Killer
Natural Pyrethrin
Sevin 5% Dust
Sevin 50 Wettable
Sevin Granules
Sevin Liquid
Sevin Spray/Liquid
Thiodan
Thiodan Vegetable & Ornamental
  Dust
Thuricide
Tomato & Veg. Fogger
Tomato & Vegetable Insect Killer
Tomato & Veg. RTU
Tomato Pepper Veg. Spray RTU
Vegetable Insect Control
Vegetable Insect Killer
Worm Ender
Yard & Garden Insect Killer

KENNELS & DOG HOUSES
Dursban Lawn Insect Spray
Flea B Gon Pet Flea & Tick Killer
Sevin 5 Dust
Sevin 10 Dust
Sevin Garden Dust

KIWI
Worm Ender

KOHLRABI
Ant Flea & Tick Insect Granules
Bioneem
Dursban 1% Granules
Dursban Ant & Turf
Dursban Lawn & Perimeter
  Granules
Flea & Tick Granules
Japanese Beetle Repellent
Liquid Flowable Sevin
Sevin 5% Dust
Sevin 10% Dust
Sevin 5 Dust
Sevin 10 Dust
Sevin Garden Dust
Sevin Granules
Sevin Liquid
Sevin Spray/Liquid
Worm Ender

LANTANA
Intercept RFO
Systemic Rose & Flower Food

55

INSECTICIDES

LARKSPUR
Lindane Borer Spray

LAURELS
Diazinon Spray
Home Patrol Insect Killer
Intercept RFO
Mite Beater
Sevin 5 Dust
Sevin 10 Dust

LEEK
Malathion 50 Plus
Worm Ender

LENTILS
Bioneem
Japanese Beetle Repellent
Worm Ender

LETTUCE
5% Diazinon Granules
5% Sevin Garden Dust
25% Diazinon Spray
50% Malathion
Bacillus thuringiensis
Bioneem
Bio-Worm Killer
Blue Dragon Garden Dust
Bug-B-Gon
Cutworm Earwig & Sowbug Bait
Cygon 2E
Diazinon 2% Granules
Diazinon 5G
Diazinon 5% G
Diazinon 5% Granules
Diazinon 12 1/2% E
Diazinon 25%
Diazinon 25% EC
Diazinon Dust
Diazinon Granules
Diazinon Plus
Diazinon Soil & Turf Insect Control
Diazinon Spray
Dipel
Dipel Dust
Earwig Sowbug & Grasshopper Bait
Flower Garden Insecticidal Soap
Fruit & Berry Insect Spray
Fruit & Veg Insect Spray
Garden Insect Spray
Garden Soil Insecticide
Garden Spray
Grasshopper Earwig & Sowbug Bait
House Plant & Garden Insect Spray
Insecticidal Soap Multi Purpose Insect Killer
Intercept Vegetable & Garden Spray
Japanese Beetle Killer
Japanese Beetle Repellent
Liquid Flowable Sevin
Liquid Sevin
Malathion 50%
Malathion 50 Plus
Many Purpose Insect Killer
Mole Cricket Killer
Natural Pyrethrin
Organic Greenhouse House & Veg. Spray
Roses & Flower Insect Spray
Saf-T-Side
Sevin 5% Dust
Sevin 10 Dust
Sevin 50% WP
Sevin Dura-Dust
Sevin Dura-Spray
Sevin Garden Dust
Sevin Granules
Sevin Liquid
Sevin Spray/Liquid
Sun Spray Ultra Fine
Systemic Granules
Thiodan Insect Spray
Thiodan Vegetable & Ornamental Dust
Thuricide
Tomato & Veg. Dust
Tomato & Veg. Fogger
Tomato & Vegetable Insect Killer
Tomato & Veg. RTU
Tomato Pepper Veg. Spray RTU
Vegetable Insect Control
Vegetable Insect Killer
Worm Ender
Yard & Garden Insect Killer

LILAC
5% Diazinon Dust
25% Diazinon
Ant Killer Powder
Bagworm & Mite Spray
Borer Killer II
Borer Miner Killer 5
Diazinon 4E
Diazinon 25%
Diazinon 25% EC
Diazinon Plus
Diazinon Spray
Dormant Oil Spray
Fruits & Vegetable Insect Spray
Garden Insect Spray
Horticultural Spray Oil
House Plant & Garden Insect Spray
Intercept H&G
Intercept RFP
Intercept Vegetable & Garden Spray
Lindane Borer Spray
Liquid Sevin
Permethrin T/O
Rose & Flower Insect Spray
Sevin 5 Dust
Sevin 10 Dust
Sevin 50% WP
Systemic House Plant Insect Control
Systemic Rose & Flower Food

LILY
Intercept Vegetable & Garden Spray
Saf-T-Side
Systemic Rose & Flower Food

LILY OF THE VALLEY
Intercept H&G

LINDEN
Dormant Oil Spray
Isotox
Orthene

LOCUST
2% Systemic Granules
25% Diazinon
Bagworm & Mite Spray
Borer Miner Killer 5
Diazinon 4E
Diazinon 12 1/2% E
Diazinon 25%
Diazinon25% EC
Diazinon Plus
Imidan
Intercept RFO
Sevin 5 Dust
Sevin 10 Dust
Systemic Granules
Systemic House Plant Insect Control

LOQUAT
Bug Buster-O
Intercept H&G

MACADAMIAS
Bioneem
Japanese Beetle Repellent
Saf-T-Side

MAGNOLIA
Dormant Spray & Summer Spray
Horticultural Spray Oil
Sevin 5 Dust
Sevin 10 Dust
Spray Oil
Systemic Rose & Flower Food

MALANGA
Worm Ender

INSECTICIDES

USAGE OF PRODUCTS

MANGO
Malathion 50 Plus
Soluable Oil Spray

MAPLE
25% Diazinon
Bagworm & Mite Spray
Borer Miner Killer 5
Diazinon 4E
Diazinon 12 1/2% E
Diazinon 25%
Diazinon 25% EC
Diazinon Plus
Diazinon Spray
Horticultural Spray Oil
Imidan
Intercept H&G
Intercept RFO
Liquid Sevin
Sevin 5 Dust
Sevin 10 Dust
Spray Oil

MARIGOLD
Aphid Mite & Whitefly Killer II
Bug-B-Gon
Bug Off Rose & Flower Spray
Fruit & Veg Insect Spray
Garden Spray
Gentle Care House Plant Spray
Home Patrol Insect Killer
House Plant & Garden
　Insect Spray
Intercept H&G
Intercept RFO
Intercept Vegetable & Garden
　Spray
Japanese Beetle Killer
Mite Beater
Multi Purpose Insect Killer
Natural Pyrethrin
Permethrin T/O
Plant Insect Control
Rose & Floral Insect Control
Rose & Floral Insect Killer
Rose & Flower Insect Killer
Roses & Flower Insect Spray
Rose & Flower Spray
Rose, Flower & Ornamental
　Insect Spray
Rose & Garden Insect Fogger
Spider Mite Control
Systemic Granules
Systemic House Plant
　Insect Control
Tomato & Veg Insect Killer
Vegetable Insect Control
Vegetable Insect Killer
Whitefly & Mealybug Control
Whitefly & Mealybug Spray

MARJORAM
Bioneem
Flower Garden Insecticidal Soap
Insecticidal Soap Multi Purpose
　Insect Killer
Japanese Beetle Repellent

MELONS
5% Diazinon Granule
5% Sevin Garden Dust
25% Diazinon
50% Malathion
Bacillus Thuringiensis
Bioneem
Bio Worm Killer
Blue Dragon Garden Dust
Bug-B-Gon
Bug Buster
Cutworm Earwig & Sowbug Bait
Cygon 2E
Diazinon 2% Granules
Diazinon 5G
Diazinon 5% G
Diazinon 5% Granules
Diazinon 12 1/2% E
Diazinon 25%
Diazinon 25% EC
Diazinon Dust
Diazinon Granules
Diazinon Plus
Diazinon Soil & Turf Insect Control
Diazinon Spray
Dipel
Dipel Dust
Earwig Sowbug & Grasshopper
　Bait
Flower Garden Insecticidal Soap
Fruit & Berry Insect Spray
Garden Insect Spray
Garden Soil Insecticide
Grasshopper Earwig & Sowbug
　Bait
Insecticidal Soap Multi Purpose
　Insect Killer
Japanese Beetle
Kelthane
Liquid Flowable Sevin
Liquid Sevin
Malathion 50%
Malathion 50 Plus
Mole Cricket Killer
Multi Purpose Insect Killer
Organic Greenhouse
　House & Veg. Spray
Saf-T-Side
Sevin 5% Dust
Sevin 10% Dust
Sevin 10 Dust
Sevin 50% WP
Sevin 50 Wettable
Sevin 50 WP

Sevin Dura-Dust
Sevin Dura-Spray
Sevin Garden Dust
Sevin Granules
Sevin Liquid
Sevin Spray/Liquid
Sun Spray Ultra Fine
Thiodan Insect Soap
Thiodan Vegetable & Ornamental
　Dust
Thuricide
Worm Ender

MIMOSA
25% Diazinon Insect Spray
Bagworm & Mite Spray
Diazinon 25%
Diazinon 25% EC
Intercept H&G
Intercept RFO

MINT
Bioneem
Japanese Beetle Repellent
Worm Ender

MOCK ORANGE
Intercept H&G
Intercept Vegetable & Garden Spray
Permethrin T/O

MOHONIA
Intercept H&G

MONSTERA
Fruit & Veg Insect Spray
House Plant & Garden
　Insect Spray
Roses & Flower Insect Spray

MORNING GLORY
Systemic Granules
Systemic House Plant
　Insect Control

MOSES IN CRADLE
Intercept RFO

MOUND TREATMENT
(Fire Ants)
5% Diazinon Granules
25% Diazinon Insect Spray
Ant Flea & Tick Insect Granules
Ant Killer Granules
Diazinon Granules
Dursban Insecticide Conc.
Dursban Lawn & Perimeter
　Granules
Fire Ant Granules
Fire Ant Killer
Primacide Fire Ant Dust
Sevin Liquid

INSECTICIDES

MOUNTAIN ASH
Horticultural Spray Oil

MULBERRY (ornamental)
Sevin 5 Dust
Sevin 10 Dust

MUSTARD
5% Diazinon Granules
5% Malathion Dust
50% Malathion
Bacillus thuringiensis
Bioneem
Blue Dragon Garden Dust
Cygon 2E
Diazinon 2% Granules
Diazinon 5G
Diazinon 5% G
Diazinon 5% Granules
Dipel
Dipel Dust
Garden Insect Spray
Garden Soil Insecticide
Garden Spray
Japanese Beetle Killer
Japanese Beetle Repellent
Liquid Flowable Sevin
Liquid Sevin
Malathion 50%
Mole Cricket Killer
Natural Pyrethrin
Sevin 5% Dust
Sevin 50 Wettable
Sevin Dura-Spray
Sevin Granules
Sevin Liquid
Sevin Spray/Liquid
Thiodan Insect Spray
Thuricide
Tomato & Veg. Fogger
Tomato & Vegetable Insect Killer
Tomato & Veg Insect Spray
Tomato & Veg. RTU
Tomato Pepper Veg. Spray RTU
Vegetable Insect Control
Vegetable Insect Killer

MUSTARD GREENS
Bio Worm Killer
Diazinon Soil & Turf Insect Control
Malathion 50 Plus
Sevin Liquid
Sevin Soil & Turf Insect Control

NANDINA
Intercept H&G

NANNYBERRY
Intercept Vegetable & Garden Spray
Permethrin T/O

NAPA CABBAGE
Worm Ender

NASTURTIUM
Intercept H&G
Intercept RFO
Sevin 5 Dust
Sevin 10 Dust
Spider Mite & Mealybug Control
Whitefly & Mealybug Spray

NECTARINES
25% Diazinon Spray
Bug-B-Gon
Bug Buster-O
Diazinon 25%
Diazinon 25% EC
Diazinon Plus
Dormant & Summer Oil Spray
Dursban Plus
Garden Insect Spray
Horticultural Spray Oil
Imidan
Liquid Flowable Sevin
Malathion 50%
Saf-T-Side
Scale Away
Sevin Dura Spray
Sevin Liquid
Sevin Spray/Liquid
Spray Oil
Summer & Dormant Oil
Sun Spray Ultra Fine
Thiodan Insect Spray
Worm Ender

NEPHTHYTIS
Aphid Mite & Whitefly Killer II
Intercept RFO
Spider Mite & Mealybug Control
Spider Mite Control
Whitefly & Mealybug Control
Whitefly & Mealybug Spray

NICOTIANA
Intercept H&G

NINEBARK
Intercept Vegetable & Garden Spray
Permethrin T/O

NUT TREES (general)
Flower Garden Insecticidal Soap
Houseplant Insecticidal Soap
Insecticidal Soap Multi Purpose Insect Killer

OAK
25% Diazinon Insect Spray
Bagworm & Mite Spray
Cygon 2E
Diazinon 4E
Diazinon 12 1/2% E
Diazinon 25%
Diazinon 25% EC
Diazinon Spray
Dormant Oil Spray
Dormant Spray & Summer Spray
Horticultural Spray Oil
Imidan
Intercept RFO
Liquid Sevin
Permethrin T/O
Sevin 5 Dust
Sevin 10 Dust
Sevin 50 WP
Spray Oil
Systemic Shrub & Flower Insecticide

OKRA
Liquid Flowable Sevin
Malathion 50 Plus
Sevin 5% Dust
Sevin 10% Dust
Sevin Dura Dust
Sevin Garden Dust
Sevin Granules
Sevin Liquid
Sevin Spray/Liquid
Worm Ender

OLEANDER
Bug-B-Gon
Intercept H&G
Multi Purpose Insect Killer
Rose & Florel Spray Bomb

OLIVES
Dormant & Summer Oil Spray
Horticultural Spray Oil
Intercept RFO
Spray Oil

ONIONS
5% Diazinon Granules
25% Diazinon
50% Malathion
Ant Flea & Tick Insect Granules
Bioneem
Bug-B-Gon
Diazinon 2% Granules
Diazinon 5G
Diazinon 5% G
Diazinon 5% Granules
Diazinon 25%
Diazinon 25% EC
Diazinon Dust
Diazinon Granules
Diazinon Plus
Diazinon Soil & Turf Insect Control
Diazinon Spray

INSECTICIDES

Dursban 1% Granules
Dursban Ant & Turf
Dursban Lawn & Perimeter
　Granules
Flea & Tick Granules
Fruit & Berry Insect Spray
Fruit & Veg Insect Spray
Garden Soil Insecticide
House Plant & Garden
　Insect Spray
Japanese Beetle Repellent
Malathion 50%
Malathion 50 Plus
Mole Cricket Killer
Organic Greenhouse
　House & Veg. Spray
Roses & Flower Insect Spray
Saf-T-Side
Worm Ender

ORCHIDS

Fruit & Veg Insect Spray
House Plant & Garden
　Insect Spray
Intercept Vegetable & Garden Spray
Permethrin T/O
Roses & Flower Insect Spray
Sevin 5 Dust
Sevin 10 Dust

ORNAMENTALS (general)

50% Malathion Insect Spray
Aphid Mite & Whitefly Killer
Bacillus thuringiensis
Bioneem
Bio Worm Killer
Blue Dragon Garden Dust
Bug Buster-O
Caterpillar Killer
Cutworm & Cricket Bait
Cutworm Earwig & Sowbug Bait
Cygon 2-E
Diazinon Dust
Dursban Ant & Turf
Dursban Granules
Dursban Insecticide
Dursban Lawn Insect Killer
Dursban Lawn & Perimeter
　Granules
Dursban Plus
Earwig Sowbug & Grasshopper
　Bait
Fire Ant Granules
Garden Insect Spray
Home Defense Home & Garden
　Insect Killer
Home Pest Control Conc.
Horticultural Spray Oil
House Plant Spray II
Insecticidal Soap Conc.
Insecticidal Soap Multi Purpose
　Insect Killer

Intercept H&G
Isotox
Japanese Beetle Killer
Japanese Beetle Repellent
Kelthane
Lindane Borer Spray
Liquid Flowable Sevin
Malathion 50%
Methoxychlor 25%
Mite & Insect Spray
Natural Pyrethrin
Organic Greenhouse
　House & Veg. Spray
Orthene
Plant Insect Control
Rose & Florel Insect Control
Rose & Florel Insect Killer
Rose & Florel Spray Bomb
Rose & Flower Dust
Rose & flower Insect Spray
Saf-T-Side
Scale Away
Sevin 5 Dust
Sevin 5% Dust
Sevin Spray/Liquid
Sevin 50 Wettable
Sevin Granules
Sevin Liquid
Sun Spray Ultra Fine
Systemic Granules
Systemic Rose Care
Systemic Rose Shrub & Flower
　Care
Thiodan Insect Spray
Thiodan Vegetable & Ornamental
　Dust
Thuricide
Volck Oil Spray
Whitefly & Mealybug Spray
Worm Ender

PACHYSANDRA

Horticultural Spray Oil
Intercept H&G

PALMS (general)

Aphid Mite & Whitefly Killer II
Dormant & Summer Oil Spray
Intercept RFO
Intercept Vegetable & Garden
　Spray
Permethrin T/O
Soluable Oil Spray
Spider Mite Control
Spider Mite & Mealybug Control
Sun Spray Ultra Fine
Whitefly & Mealybug Control
Whitefly & Mealybug Spray

PALM (Kantia)

Home Patrol Insect Killer
Mite Beater

PALM (Sago)

Dormant Spray & Summer Spray
Intercept H&G
Intercept RFO

PANSY

Bug-B-Gon
Intercept H&G
Intercept RFO
Intercept Vegetable & Garden
　Spray
Multi Purpose Insect Killer
Permethrin T/O
Spider Mite & Mealybug Control
Systemic Granules
Systemic House Plant
　Insect Control
Whitefly & Mealybug Spray

PAPAYA

Malathion 50%

PARSLEY

5% Diazinon Granules
Bioneem
Cutworm Earwig & Sowbug Bait
Diazinon 2% Granules
Diazinon 5G
Diazinon 5% G
Diazinon 5% Granules
Diazinon Granules
Diazinon Soil & Turf Insect Control
Earwig Sowbug & Grasshopper
　Bait
Garden Soil Insecticide
Grasshopper Earwig & Sowbug
　Bait
Japanese Beetle Repellent
Liquid Flowable Sevin
Liquid Sevin
Mole Cricket Killer
Sevin 5% Dust
Sevin Granules
Sevin Liquid
Sevin Spray/Liquid
Worm Ender

PARSNIPS

25% Diazinon Insect Spray
Bug-B-Gon
Cutworm Earwig & Sowbug Bait
Diazinon 25%
Diazinon 25% EC
Diazinon Dust
Diazinon Plus
Earwig Sowbug & Grasshopper
　Bait
Grasshopper Earwig & Sowbug
　Bait
Liquid Flowable Sevin
Sevin 5% Dust

USAGE OF PRODUCTS

USAGE OF PRODUCTS

INSECTICIDES

Sevin Dura Spray
Sevin Granules
Sevin Liquid
Sevin Spray/Liquid
Worm Ender

PEACHES

25% Diazinon
25% Methoxychlor Spray
50% Malathion
Bioneem
Borer Killer II
Borer Miner Killer 5
Bug-B-Gon
Bug Buster-O
Diazinon 12 1/2% E
Diazinon 25%
Diazinon 25% EC
Diazinon Plus
Diazinon Spray
Dormant Oil Spray
Dormant & Summer Oil Spray
Dormant Spray & Summer Spray
Dursban Plus
Fruit & Berry Insect Spray
Fruit & Veg Insect Spray
Garden Insect Spray
Horticultural Spray Oil
House Plant & Garden
  Insect Spray
Imidan
Intercept Vegetable & Garden
  Spray
Japanese Beetle Repellent
Liquid Flowable Sevin
Malathion 50%
Malathion 50 Plus
Methoxychlor 25%
Organic Greenhouse
  House & Veg. Spray
Permethrin T/O
Roses & Flower Insect Spray
Saf-T-Side
Scale Away
Sevin 10 Dust
Sevin 50% WP
Sevin 50 WP
Sevin Dura Spray
Sevin Liquid
Spray Oil
Summer & Dormant Oil
Sun Spray Ultra Fine
Thiodan Insect Spray
Volck Oil Spray
Worm Ender

PEANUTS

Bioneem
Dursban Ant & Turf
Japanese Beetle Repellent
Saf-T-Side
Sevin Liquid

PEARS

25% Methoxychlor Spray
50% Malathion
Bioneem
Bug Buster-O
Cutworm Earwig & Sowbug Bait
Diazinon Dust
Diazinon Granules
Diazinon Spray
Dormant Oil Spray
Dormant Spray & Summer Spray
Dormant & Summer Oil Spray
Dursban Plus
Earwig Sowbug & Grasshopper
  Bait
Fruit & Berry Insect Spray
Fruit & Veg Insect Spray
Garden Insect Spray
Grasshopper Earwig & Sowbug
  Bait
Horticultural Spray Oil
House Plant & Garden
  Insect Spray
Imidan
Intercept Vegetable & Garden
  Spray
Japanese Beetle Repellent
Kelthane
Liquid Flowable Sevin
Malathion 50%
Malathion 50 Plus
Metoxychlor 25%
Organic Greenhouse
  House & Veg. Spray
Permethrin t/O
Roses & Flower Insect Spray
Saf-T-Side
Scale Away
Sevin 50% WP
Sevin 50 WP
Sevin 50 Wettable
Sevin Liquid
Sevin Dura Spray
Spray Oil
Summer & Dormant Oil
Sun Spray Ultra Fine
Volck Oil Spray
Worm Ender

PEAS

5% Diazinon Granules
5% Malathion Dust
5% Sevin Garden Dust
25% Diazinon Spray
25% Methoxychlor Spray
50% Malathion Spray
Bioneem
Bug-B-Gon
Bug Buster
Cygon 2E
Diazinon 2% Granules
Diazinon 5G
Diazinon 5% G
Diazinon 5% Granules
Diazinon 12 1/2% E
Diazinon 25%
Diazinon25% EC
Diazinon Plus
Diazinon Soil & Turf Insect Control
Flower Garden Insecticidal Soap
Fruit & Veg Insect Spray
Garden Insect Spray
Garden Soil Insecticide
House Plant & Garden
  Insect Spray
Insecticidal Soap Multi Purpose
  Insect Killer
Japanese Beetle Repellent
Liquid Flowable Sevin
Liquid Sevin
Malathion 50%
Malathion 50 Plus
Methoxychlor 25%
Mole Cricket Killer
Multi Purpose Insect Killer
Organic Greenhouse
  House & Veg. Spray
Roses & Flower Insect Spray
Sevin 50 WP
Sevin Dura-Spray
Sevin Granules
Sevin Liquid
Sevin Spray/Liquid
Systemic Granules
Tomato & Veg. Dust
Worm Ender

PEA SHRUB

Intercept Vegetable & Garden Spray
Permethrin T/O

PECANS

50% Malathion
Bioneem
Dormant Spray & Summer Spray
Garden Insect Spray
Japanese Beetle Repellent
Liquid Flowable Sevin
Malathion 50%
Saf-T-Side
Scale Away
Sevin Dura Spray
Sevin Liquid
Sevin Spray/Liquid
Summer & Dormant Oil
Sun Spray Ultra Fine
Thiodan Insect Spray
Worm Ender

PEONY

Intercept RFO
Rose & Florel Spray Bomb
Systemic Rose & Flower Food

INSECTICIDES

PEPEROMIA
- Aphid Mite & Whitefly Killer
- Aphid Mite & Whitefly Killer II
- Intercept RFO
- Intercept Vegetable & Garden Spray
- Spider Mite Control
- Spider Mite & Mealbug Control
- Whitefly & Mealybug Control
- Whitefly & Mealybug Spray

PEPPERS
- 5% Diazinon Granules
- 5% Malathion Dust
- 25% Diazinon
- 25% methoxychlor Spray
- 50% Malathion
- Bioneem
- Blue Dragon Garden Dust
- Bug-B-Gon
- Bug Buster
- Cutworm Earwig & Sowbug Bait
- Cygon 2E
- Diazinon 2% Granules
- Diazinon 5G
- Diazinon 5% G
- Diazinon 5% Granules
- Diazinon 12 1/2% E
- Diazinon 25%
- Diazinon 25% EC
- Diazinon Dust
- Diazinon Granules
- Diazinon Plus
- Diazinon Soil & Turf Insect Control
- Earwig Sowbug & Grasshopper Bait
- Flower Garden Insecticidal Soap
- Fruit & Veg Insect Spray
- Garden Insect Spray
- Garden Soil Insecticide
- Garden Spray
- Grasshopper Earwig & Sowbug Bait
- House Plant & Garden Insect Spray
- Insecticidal Soap Multi Purpose Insect Killer
- Intercept Vegetable & Garden Spray
- Japanese Beetle Killer
- Japanese Beetle Repellent
- Kelthane
- Liquid Flowable Sevin
- Liquid Sevin
- Malathion 50%
- Malathion 50 Plus
- Many Purpose Insect Killer
- Mole Cricket Killer
- Multipurpose Insect Killer
- Natural Pyrethrin
- Organic Greenhouse House & Veg. Spray
- Permethrin T/O
- Red Spider Spray
- Roses & Flower Insect Spray
- Saf-T-Side
- Sevin 5 Dust
- Sevin 10 Dust
- Sevin 5% Dust
- Sevin 10% Dust
- Sevin Garden Dust
- Sevin Granules
- Sevin Liquid
- Sevin Wettable
- Sevin 50 WP
- Sevin Dura-Dust
- Sevin Dura-Spray
- Sevin Spray/Liquid
- Sun Spray Ultra Fine
- Thiodan Insect Spray
- Thiodan Vegetable & Ornamental Dust
- Tomato & Veg. Fogger
- Tomato & Veg Insect Killer
- Tomato & Veg Insect Spray
- Tomato & Veg. RTU
- Tomato Pepper Veg. Spray RTU
- Vegetable Insect Control
- Vegetable Insect Killer
- Worm Ender

PEPPERONIA
- Fruits & Vegetable Insect Spray
- House Plant & Garden Insect Spray
- Roses & Flower Insect Spray
- Rose & Flower Insect Killer II

PERSIMMONS
- Worm Ender

PETUNIA
- 2% Systemic Granules
- Bug-B-Gon
- Fruit & Veg Insect Spray
- House Plant & Garden Insect Spray
- Intercept H&G
- Intercept RFO
- Intercept Vegetable & Garden Spray
- Multi Purpose Insect Killer
- Permethrin T/O
- Roses & Flower Insect Spray
- Sevin 5 Dust
- Sevin 10 Dust
- Spider Mite & Mealybug Control
- Systemic Granules
- Systemic House Plant Insect Control
- Whitefly & Mealybug Spray

PHILODENDRON
- Aphid Mite & Whitefly Killer
- Aphid Mite & Whitefly Killer II
- Fruits & Vegetable Insect Spray
- House Plant & Garden Insect Spray
- Intercept RFO
- Intercept Vegetable & Garden Spray
- Permethrin T/O
- Rose & Flower Insect Killer II
- Roses & Flowers Insect Spray
- Spider Mite Control
- Spider Mite & Mealybug Control
- Sun Spray Ultra Fine
- Systemic Rose & Flower Food
- Whitefly & Mealybug Control
- Whitefly & Mealybug Spray

PHOTINIA
- Intercept H&G

PICEA
- Intercept H&G

PIERIS
- Intercept H&G

PIGGY BACK PLANT
- Aphid Mite & Whitefly Killer II
- Intercept RFO
- Intercept Vegetable & Garden Spray
- Spider Mite Control
- Spider Mite & Mealybug Control
- Whitefly & Mealybug Control
- Whitefly & Mealybug Spray

PINES
- 2% Systemic Granules
- 25% Diazinon
- Bagworm & Mite Spray
- Borer Miner Killer 5
- Cygon 2E
- Delpheniums
- Diazinon 4E
- Diazinon 12 1/2% E
- Diazinon 25%
- Diazinon 25% EC
- Diazinon Plus
- Diazinon Spray
- Dormant Spray & Summer Spray
- Garden Insect Spray
- Home Patrol Insect Killer
- Horticultural Spray Oil
- Imidan
- Intercept H&G
- Lindane Borer Spray
- Liquid Sevin
- Permethrin T/O
- Sevin 5 Dust

USAGE OF PRODUCTS

61

INSECTICIDES

Sevin 10 Dust
Systemic Granules
Systemic House Plant
   Insect Control
Systemic Shrub & Flower
   Insecticide
Thiodan Insect Spray

PINEAPPLE

Malathion 50%
Worm Ender

PISTACHIOS

Bioneem
Japanese Beetle Repellent

PITTOSPORUM

Intercept RFO
Soluable Oil Spray

PLUMS

25% Diazinon
25% Methoxychlor Spray
50% Malathion
Bagworm & Mite Spray
Bioneem
Borer Miner Killer 5
Bug-B-Gon
Bug Buster-O
Diazinon 4E
Diazinon 12 1/2% E
Diazinon 25%
Diazinon 25% EC
Diazinon Plus
Diazinon Spray
Dormant Oil Spray
Dormant & Summer Spray Oil
Dormant Spray & Summer Spray
Dursban Plus
Fruit & Berry Insect Spray
Fruit & Veg Insect Spray
Garden Insect Spray
Horticultural Spray Oil
House Plant & Garden
   Insect Spray
Imidan
Japanese Beetle Repellent
Liquid Flowable Sevin
Malathion 50%
Metoxychlor 25%
Organic Greenhouse
   House & Veg. Spray
Roses & Flower Insect Spray
Scale Away
Sevin 10 Dust
Sevin 50% WP
Sevin 50 WP
Sevin Dura Spray
Sevin Liquid
Spray Oil
Summer & Dormant Oil
Sun Spray Ultra Fine

Thiodan Insect Spray
Volck Oil Spray
Worm Ender

PLUMS (ornamental)

Diazinon 25%

PODOCARPUS

Intercept RFO

POINSETTIA

Cygon 2E
Intercept H&G
Intercept RFO
Intercept Vegetable & Garden
   Spray
Permethrin T/O
Saf-T-Side
Spider Mite & Mealybug Control
Sun Spray Ultra Fine
Systemic Shrub & Flower
   Insecticide
Whitefly & Mealybug Spray

POLKA DOT PLANT

Aphid Mite & Whitefly Killer II
Spider Mite Control
Whitefly & Mealybug Control

POMEGRANATE

Intercept H&G
Worm Ender

POPCORN

Saf-T-Side

POPLAR

25% Diazinon Insect Spray
Bagworm & Mite Spray
Borer Miner Killer 5
Diazinon 4E
Diazinon 25%
Diazinon 25% EC
Intercept H&G
Sevin 5 Dust
Sevin 10 Dust

PORTULACA

Intercept H&G
Intercept Vegetable & Garden
   Spray
Sun Spray Ultra Fine

POTATOES

5% Diazinon Granules
5% Sevin Garden Dust
25% Diazinon
25% Methoxychlor Insect Spray
50% Malathion
Bacillus thuringiensis
Bioneem
Blue Dragon Garden Dust

Bug-B-Gon
Bug Buster
Colo. Potato Beetle Beater
Cygon 2E
Diazinon 2% Granules
Diazinon 5G
Diazinon 5% G
Diazinon 5% Granules
Diazinon 12 1/2% E
Diazinon 25%
Diazinon 25% EC
Diazinon Dust
Diazinon Granules
Diazinon Plus
Diazinon Soil & Turf Insect Control
Diazinon Spray
Dipel
Dipel Dust
Flower Garden Insecticidal Soap
Fruit & Berry Insect Spray
Fruit & Veg Insect Spray
Garden Insect Spray
Garden Soil Insecticide
Garden Spray
House Plant & Garden
   Insect Spray
Insecticidal Soap Multi Purpose
   Insect Killer
Intercept Vegetable & Garden
   Spray
Japanese Beetle Killer
Japanese Beetle Repellent
Liquid Flowable Sevin
Liquid Sevin
Malathion 50%
Malathion 50 Plus
Methoxychlor 25%
Mole Cricket Killer
Natural Pyrethrin
Organic Greenhouse
   House & Veg. Spray
Permethrin T/O
Roses & Flower Insect Spray
Saf-T-Side
Sevin 5 Dust
Sevin 10 Dust
Sevin 5% Dust
Sevin 10% Dust
Sevin 50 WP
Sevin Dura-Dust
Sevin Dura-Spray
Sevin Garden Dust
Sevin Granules
Sevin Liquid
Sevin Spray/Liquid
Sun Spray Ultra Fine
Systemic Granules
Thiodan Insect Spray
Thiodan Vegetable & Oranametal
   Dust
Tomato & Veg. Fogger
Tomato & Vegetable Insect Killer

INSECTICIDES

Tomato & Garden
Tomato & Veg. RTU Insect Killer
Vegetable Insect Control
Vegetable Insect Killer
Worm Ender
Yard & Garden Insect Killer

POTHOS

Aphid Mite & Whitefly Killer II
Intercept RFO
Intercept Vegetable & Garden Spray
Spider Mite Control
Spider Mite & Mealybug Control
Whitefly & Mealybug Control
Whitefly & Mealybug Spray

POTTED PLANTS (general)

2% Systemic Granules
Rose & Floral Spray Bomb
Systemic Granules
Systemic Rose Care
Systemic Rose & Flower Care
Systemic Rose Shrub & Flower Care
Systemic Rose & Flower Food

POULTRY (houses)

Sevin 5 Dust
Sevin 10 Dust

POULTRY (non commercial)

5% Malathion Dust
Livestock Dust
Sevin 5 Dust
Sevin 10 Dust

PRAYER PLANT

Aphid Mite & Whitefly Killer II
Intercept RFO
Intercept Vegetable & Garden Spray
Spider Mite Control
Spider Mite & Mealybug Control
Whitefly & Mealybug Control
Whitefly & Mealybug Spray

PREMISES (general)

5% Diazinon Granules
25% Diazinon
25% Methoxychlor Insect Spray
50% Malathion
Ant Barrier
Ant Flea & Tick Insect Granules
Ant, Flea & Spider Killer
Ant Flea & Tick Killer
Ant Killer Dust II
Ant Killer Granules
Ant Killer Plus
Ant Roach & Spider Control
Ant Stop Bait 2
Ant Stop Dust

Bioneem
Bug-B-Gon
Bug Buster
Carpenter Ant Killer
Cygon 2E
Diazinon 2% Granules
Diazinon 5% Granules
Diazinon 25%
Diazinon 25% EC
Diazinon Granules
Diazinon Plus
Diazinon Soil & Turf Insect Control
Diazinon Spray
Dursban Ant & Turf
Dursban Granules
Dursban Insecticide
Dursban Insecticide Conc.
Dursban Insect Spray
Dursban Lawn Insect Killer
Dursban Lawn Insect Spray
Dursban Lawn Insect Granular
Dursban Lawn & Perimeter Granules
Dursban Plus
Dursban Ready Spray
Fire Ant Granules
Fire Ant Killer
Fire Ant Killer Granules
Flea & Tick Spray
Flea B Gon Flea & Tick Killer
Flea B gon Pet Flea & Tick Killer
Flea B Gon Outdoor Flea & Tick Killer
Home Defense Home & Garden Insect Killer
Home Defense Indoor Outdoor Insect Killer
Home Patrol Insect Killer
Home Pest Control
Hornet & Wasp Killer 2
Intercept H&G
Intercept RFO
Japanese Beetle Repellent
Logic Fire Ant Bait
Malathion 50%
Malathion 50 Plus
Mite & Insect Spray
Mosquito Larvicide
Mosquito Spray Conc.
Natural Pyrethrin
Oftanol
Organic Greenhouse House & Veg. Spray
Orthene
Orthene Fire Ant Killer
Ortho-Klor
Outdoor Insect Fogger
Roach-Rid Home Pest Killer
Sevin 5% Dust
Sevin Liquid
Superstop Ant Roach & Insect Killer
Termi-Chlor

Termite Killer
Ultra S.S.C.
Vikor Concentrate
Vikor Dual Action
Vikor Ready to Use
Wasp & Hornet Jet Spray
Wasp & Hornet Killer
Wasp & Hornet Spray
Whack Hornet & Wasp Killer

PRIMROSE

Mite & Insect Spray

PRIVET (Ligustrum)

Dormant Spray & Summer Spray
Horticultural Spray Oil
Intercept H&G
Malathion-Oil
Sevin 5 Dust
Sevin 10 Dust
Soluable Oil Spray
Spray Oil

PRUNES

25% Diazinon Insect Spray
25% Methoxychlor Insect Spray
50% Malathion
Bug-B-Gon
Bug Buster-O
Diazinon 12 1/2% E
Diazinon 25%
Diazinon Plus
Dormant & Summer Spray Oil
Dursban Plus
Garden Insect Spray
Horticultural Spray Oil
Imidan
Liquid Flowable Sevin
Malathion 50%
Methoxychlor 25%
Scale Away
Sevin 10 Dust
Sevin Liquid
Spray Oil
Summer & Dormant Oil
Worm Ender

PRUNUS SPP. (general)

Intercept H&G

PSEUDOTSUGA

Intercept H&G

PUMPKINS

5% Sevin Garden Dust
Bioneem
Blue Dragon Garden Dust
Bug Buster
Garden Insect Spray
Insecticidal Soap Multi Purpose Insect Killer
Kelthane

USAGE OF PRODUCTS

63

INSECTICIDES

Liquid Flowable Sevin
Liquid Sevin
Malathion 50 Plus
Organic Greenhouse
   House & Veg. Spray
Saf-T-Side
Sevin 5% Dust
Sevin 10% Dust
Sevin 10 Dust
Sevin 50 Wettable
Sevin Garden Dust
Sevin Granules
Sevin Liquid
Sevin Dura Spray
Sevin Spray/Liquid
Thiodan Insect Spray
Thiodan Vegetable & Ornamental Dust
Worm Ender

PURPLE PASSION

Aphid Mite & Whitefly Killer II
Intercept Vegetable & Garden Spray
Rose & Florel Spray Bait
Spider Mite Control
Whitefly & Mealybug Control

PURPLE WAFFLES

Intercept RFO
Spider Mite & Mealybug Control
Whitefly & Mealybug Spray

PYRACANTHA

2% Systemic Granules
5% Malathion Dust
Bug-B-Gon
Diazinon Plus
Horticultural Spray Oil
Intercept H&G
Lindane Borer Spray
Multi Purpose Insect Killer
Soluable Oil Spray
Spray Oil
Systemic Granules
Systemic House Plant Insect Control
Systemic Granules

PYRUS

Intercept H&G

QUINCE

Bug Buster-O
Dormant Oil Spray
Kelthane
Organic Greenhouse
   House & Veg. Spray
Systemic Rose & Flower Food
Worm Ender

RADISHES

5% Diazinon Granules
25% Diazinon Spray
50% Malathion
Bioneem
Bug-B-Gon
Bug Buster
Cutworm Earwig & Sowbug Bait
Diazinon 2% Granules
Diazinon 5G
Diazinon 5% G
Diazinon 5% Granules
Diazinon 12 1/2% E
Diazinon 25%
Diazinon 25% EC
Diazinon Dust
Diazinon Plus
Diazinon Spray
Dursban 1% Granules
Dursban Ant & Turf
Dursban Granules
Dursban Lawn & Perimeter Granules
Earwig Sowbug & Grasshopper Bait
Flea & Tick Granules
Flower Garden Insecticidal Soap
Fruit & Berry Insect Spray
Fruit & Veg Insect Spray
Garden Soil Insecticide
Garden Spray
Grasshopper Earwig & Sowbug Bait
House Plant & Garden Insect Spray
Insecticidal Soap Multi Purpose Insect Killer
Japanese Beetle Killer
Japanese Beetle Repellent
Liquid Flowable Sevin
Liquid Sevin
Malathion 50%
Malathion 50 Plus
Mole Cricket Killer
Natural Pyrethrin
Organic Greenhouse
   House & Veg. Spray
Roses & Flowers Insect Spray
Saf-T-Side
Sevin 5% Dust
Sevin Dura-Spray
Sevin Granules
Sevin Liquid
Sevin Spray/Liquid
Sun Spray Ultra Fine
Tomato & Veg. Fogger
Tomato & Vegetable Insect Killer
Tomato & Vegetable Insect Spray
Tomato & Veg. RTU
Tomato Pepper Veg. Spray RTU
Vegetable Insect Control
Vegetable Insect Killer
Worm Ender

REDBUD

Horticultural Spray Oil

RHODODENDRON

2% Systemic Granules
5% Diazinon Dust
25% Diazinon
Ant Killer Powder
Bagworm & Mite Spray
Borer Miner Killer 5
Diazinon 4E
Diazinon 25%
Diazinon 25% EC
Diazinon Plus
Diazinon Spray
Home Patrol Insect Killer
Horticultural Spray Oil
Intercept H&G
Intercept RFO
Intercept Vegetable & Garden Spray
Lindane Borer Spray
Mite Beater
Mite & Insect Spray
Multi Purpose Insect Killer
Permethrin T/O
Pestkil
Saf-T-Side
Sevin 5 Dust
Sevin 10 Dust
Systemic Granules
Systemic House Plant Insect Control
Systemic Rose & Flower Food

RIBBON PLANT

Aphid Mite & Whitefly Killer II
Spider Mite Control
Whitefly & Mealybug Control

ROSE

2% Systemic Granules
5% Diazinon Dust
5% Malathion Dust
25% Diazinon
50% Malathion
Ant Killer Powder
Aphid Mite & Whitefly Killer
Bagworm & Mite Spray
Blue Dragon Garden Dust
Bug-B-Gon
Bug Off Rose & Flower Spray
Cygon 2E
Diazinon 4E
Diazinon 12 1/2% E
Diazinon 25%
Diazinon 25% EC
Diazinon Dust
Diazinon Plus

INSECTICIDES

USAGE OF PRODUCTS

Diazinon Spray
Dormant Oil Spray
Dormant Spray & Summer Spray
Feed & Shield Systemic Rose Care
Flower Garden Insecticidal Soap
Fruit & Veg Insect Spray
Garden Spray
Gentle Care House Plant Spray
Home Patrol Insect Killer
Horticultural Spray Oil
House Plant & Garden Insect Spray
Insecticidal Soap Multi Purpose Insect Killer
Intercept H&G
Intercept RFO
Intercept Vegetable & Garden Spray
Isotox
Japanese Beetle Killer
Lindane Borer Spray
Liquid Sevin
Malathion 50 Plus
Mite Beater
Multi Purpose Insect Killer
Natural Pyrethrin
Orthene
Permethrin T/O
Pestkil
Plant Insect Control
Rose & Floral Insect Control
Rose & Floral Insect Killer
Rose & Flower Care
Rose & Flower Dust
Rose & Flower Insect Killer
Rose & Flower Insect Killer II
Rose & Flower Insect Spray
Rose & Flower Spray
Rose & Flower Spray Bomb
Rose, Flower & Ornamental Insect Spray
Rose & Garden Insect Fogger
Rose Guard
Saf-T-Side
Sevin 5% Dust
Sevin 50% WP
Sevin 50 WP
Sevin 50 Wettable
Sevin Dura-Spray
Sevin Liquid
Sevin Spray/Liquid
Soluable Oil Spray
Systemic Granules
Systemic House Plant Insect Control
Systemic Rose & Flower Care
Systemic Rose & Flower Food
Tomato & Vegetable Insect Killer
Vegetable Insect Control
Vegetable Insect Killer
Volck Oil Spray
Whitefly & Mealybug Spray

ROSE OF SHARON (Athea)
Intercept RFO
Sevin 5 Dust
Sevin 10 Dust

ROSEMARY
Bioneem
Japanese Beetle Repellent

RUBBER PLANT
Aphid Mite & Whitefly Killer II
Fruit & Veg Insect Spray
Gentle Care House Plant Spray
Home Patrol Insect Killer
House Plant & Garden Insect Spray
Intercept RFO
Mite Beater
Natural Pyrethrin
Roses & Flower Insect Spray
Rose & Garden Insect Fogger
Spider Mite Control
Spider Mite & Mealybug Cotnrol
Whitefly & Mealybug Control
Whitefly & Mealybug Spray

RUTABAGA
Ant Flea & Tick Insect Granules
Bioneem
Cutworm Earwig & Sowbug Bait
Dursban 1% Granules
Dursban Granules
Dursban Lawn & Perimeter Granules
Earwig Sowbug & Grasshopper Bait
Flea & Tick Granules
Grasshopper Earwig & Sowbug Bait
Japanese Beelte Repellent
Liquid Flowable Sevin
Liquid Sevin
Malathion 50 Plus
Sevin 5% Dust
Sevin Dura Spray
Sevin Granules
Sevin Liquid
Sevin Spray/Liquid
Worm Ender

SAGE
Bioneem
Flower Garden Insecticidal Soap
Insecticidal Soap Multi Purpose Insect Killer
Japanese Beetle Repellent

SALSIFY
Liquid Flowable Sevin
Liquid Sevin
Sevin 5% Dust

Sevin Granules
Sevin Liquid
Sevin Spray/Liquid
Worm Ender

SALVIA
Aphid Mite & Whitefly Killer II
Bug-B-Gon
Fruit & Veg Insect Spray
House Plant & Garden Insect Spray
Intercept H&G
Intercept RFO
Multi Purpose Insect Killer
Roses & Flower Insect Spray
Spider Mite Control
Whitefly & Mealybug Control
Whitefly & Mealybug Spray

SANSEWIERIA
Aphid Mite & Whitefly Killer II
Spider Mite Control
Whitefly & Mealybug Control

SASSAFRAS
Sevin 5 Dust
Sevin 10 Dust

SCHEFFLERA (Brassaia)
Aphid Mite & Whitefly Killer II
Intercept H&G
Intercept Vegetable & Garden Spray
Spider Mite Control
Systemic Rose & Flower Food
Whitefly & Mealybug Control

SEDUM
Aphid Mite & Whitefly Killer
Intercept RFO
Rose & Flower Insect Killer II
Whitefly & Mealybug Spray

SESAME
Bug Buster-O

SHALLOTS
Malathion 50 Plus
Worm Ender

SHAMROCK
Intercept RFO
Whitefly & Mealybug Spray

SHRUBS (general)
2% Systemic Granules
50% Malathion
Bagworm & Mite Spray
Bioneem
Bug-B-Gon
Caterpillar Killer
Cutworm Earwig & Sowbug Bait

65

INSECTICIDES

Diazinon 25% EC
Diazinon Dust
Dormant Spray
Dormant & Summer Oil Spray
Dursban
Dursban Ant & Turf
Dursban Insecticide
Dursban Insecticide Conc.
Dursban Lawn Insect Killer
Earwig Sowbug & Grasshopper Bait
Flea & Tick Granules
Fruits & Vegetable Insect Spray
Horticultural Spray Oil
Houseplant & Garden Insect Spray
Houseplant Insecticidal Soap
Insecticidal Soap Multi Purpose Insect Killer
Isotox
Japanese Beetle Killer
Kelthane
Lindane Berer Spray
Liquid Flowable Sevin
Liquid Sevin
Malathion 50%
Many Purpose Dursban Concentrate
Japanese Beetle Repellent
Mite & Insect Spray
Multi Purpose Insect Killer
Orthene
Pestkil
Rose & Flower Care
Roses & Flower Insect Spray
Rose & Flower Insect Killer II
Rose & Flower Insect Spray
Rose, Flower & Ornamental Insect Spray
Scale Away
Sevin 50 Wettable
Sevin Dura-Spray
Sevin Liquid
Sevin Spray/Liquid
Spray Oil
Summer & Dormant Oil
Sun Spray Ultra Fine
Systemic Granules
Systemic Rose Care
Systemic Rose & Flower Care
Systemic Rose & Flower Food
Systemic Rose Shrub & Flower Care
Thiodan Insect Spray
Tomato & Vegetable Insect Killer
Volck Oil Spray
Whack Hornet & Wasp Killer
Yard & Garden Insect Killer

SILVER TREE

Aphid Mite & Whitefly Killer II
Spider Mite Control
Whitefly & Mealybug Control

SMALL FRUITS/BERRIES (general)

Bug Buster-O

SNAKE PLANT

Intercept RFO
Intercept Vegetable & Garden Spray

SNAPDRAGON

Bug-B-Gon
Home Patrol Insect Killer
Intercept H&G
Intercept RFO
Intercept Vegetable & Garden Spray
Mite Beater
Multi Purpose Insect Killer
Permethrin T/O
Whitefly & Mealybug Spray

SNOWBERRY

Intercept Vegetable & Garden Spray
Permethrin T/O

SPIDER PLANT

Aphid Mite & Whitefly Killer II
Spider Mite Control
Whitefly & Mealbug Control

SPINACH

5% Diazinon Granules
Bacillus thuringiensis
Bioneem
Bug-B-Gon
Cutworm Earwig & Sowbug Bait
Diazinon 2% Granules
Diazinon 5G
Diazinon 5% G
Diazinon 25%
Diazinon 25% EC
Diazinon Granules
Diazinon Plus
Diazinon Soil & Turf Insect Control
Dipel
Dipel Dust
Earwig Sowbug & Grasshopper Bait
Flower Garden Insecticidal Soap
Garden Insect Spray
Garden Soil Insecticide
Garden Spray
Grasshopper Earwig & Sowbug Bait
Insecticidal Soap Multi Purpose Insect Killer
Intercept Vegetable & Garden Spray
Japanese Beetle Killer
Japanese Beetle Repellent
Liquid Flowable Sevin
Malathion 50 Plus
Many Purpose Insect Killer
Mole Cricket Killer
Natural Pyrethrin
Organic Greenhouse House & Veg. Spray
Permethrin T/O
Sevin 5% Dust
Sevin 50 Wettable
Sevin Dura-Spray
Sevin Granules
Sevin Liquid
Sevin Spray/Liquid
Thiodan Vegetable & Ornamental Dust
Tomato & Veg. Dust
Tomato & Vegetable Insect Killer
Tomato & Veg. RTU
Tomato Pepper Veg. Spray RTU
Thuricide
Vegetable Insect Control
Vegetable Insect Killer
Worm Ender

SPIRIA

2% Systemic Granules
Aphid Mite & Whitefly Killer
Intercept H&G
Intercept RFO
Rose & Florel Spray Bomb
Rose & Flower Insect Killer II
Sevin 5 Dust
Sevin 10 Dust
Systemic Granules
Systemic House Plant Insect Control
Systemic Rose & Flower Food

SPRUCE

25% Diazinon
Bagworm & Mite Spray
Borer Miner Killer 5
Diazinon 4E
Diazinon 12 1/2% E
Diazinon 25%
Diazinon 25% EC
Diazinon Plus
Diazinon Spray
Horticultural Spray Oil
Imidan
Intercept RFO

SQUASH

5% Diazinon Granules
5% Malathion Dust
5% Sevin Garden Dust
25% Diazinon Spray
50% Malathion
Bioneem
Blue Dragon Garden Dust
Bug-B-Gon

INSECTICIDES

Bug Buster
Cutworm Earwig & Sowbug Bait
Diazinon 2% Granules
Diazinon 5G
Diazinon 5% G
Diazinon 5% Granules
Diazinon 12 1/2% E
Diazinon 25%
Diazinon 25% EC
Diazinon Dust
Diazinon Granules
Diazinon Plus
Diazinon Soil & Turf Insect Control
Diazinon Spray
Earwig Sowbug & Grasshopper Bait
Flower Garden Insecticidal Soap
Fruit & Berry Insect Spray
Fruit & Veg Insect Spray
Garden Insect Spray
Garden Soil Insecticide
Grasshopper Earwig & Sowbug Bait
House Plant & Garden Insect Spray
Insecticidal Soap Multi Purpose Insect Killer
Japanese Beetle Repellent
Liquid Flowable Sevin
Liquid Sevin
Malathion 50%
Malathion 50 Plus
Many Purpose Insect Killer
Mole Cricket Killer
Multi Purpose Insect Killer
Organic Greenhouse House & Veg. Spray
Roses & Flower Insect Spray
Saf-T-Side
Sevin 5 Dust
Sevin 5% Dust
Sevin 10% Dust
Sevin 10 Dust
Sevin Garden Dust
Sevin Liquid
Sevin Wettable
Sevin 50% WP
Sevin 50 WP
Sevin Dura-Dust
Sevin Dura-Spray
Sevin Spray/Liquid
Sun Spray Ultra Fine
Thiodan Insect Spray
Thiodan Vegetable & Ornamental Dust
Tomato & Vegetable Insect Killer
Worm Ender
Yard & Garden Insect Killer

STAR OF BETHLEHEM

Intercept RFO
Whitefly & Mealybug Spray

STATICE

Intercept Vegetable & Garden Spray

STOCKS

50% Malathion Insect Spray
Home Patrol Insect Killer
Malathion 50 Plus

STRAWBERRIES

5% Malathion Dust
25% Diazinon Spray
25% Methoxychlor Spray
50% Malathion
Bug-B-Gon
Cutworm Earwig & Sowbug Bait
Diazinon 12 1/2% E
Diazinon 25%
Diazinon Dust
Diazinon Plus
Diazinon Spray
Earwig Sowbug & Grasshopper Bait
Fruit & Berry Insect Spray
Fruit & Veg Insect Spray
Garden Insect Spray
Garden Soil Insecticide
Grasshopper Earwig & Sowbug Bait
House Plant & Garden Insect Spray
Kelthane
Liquid Flowable Sevin
Malathion 50%
Malathion 50 Plus
Methoxychlor 25%
Mite Beater
Mole Cricket Killer
Multi Purpose Insect Killer
Organic Greenhouse House & Veg. Spray
Red Spider Spray
Roses & Flower Insect Spray
Saf-T-Side
Sevin 5 Dust
Sevin 10 Dust
Sevin 5% Dust
Sevin 50 Wettable
Sevin Garden Dust
Sevin Granules
Sevin Liquid
Sevin Spray/Liquid
Sun Spray Ultra Fine
Thiodan Insect Spray
Thiodan Vegetable & Ornamental Dust
Tomato & Vegetable Dust

SUB-TROPICAL FRUITS (general)

Bug Buster-O

SUCCULANTS (general)

Flower Garden Insecticidal Soap
Fruit & Veg Insect Spray
House Plant & Garden Insect Spray
HousePlant Insecticidal Soap
Insecticidal Soap Multi Purpose Insect Killer
Roses & Flower Insect Spray

SUGAR BEETS

Worm Ender

SUNFLOWER

Bug Buster-O
Dursban Ant & Turf
Dursban Lawn & Perimeter Granules
Garden Insect Spray

SWEET BAY

Bioneem
Japanese Beetle Repellent

SWEET PEAS

Rose & Flower Dust

SWEET POTATOES

5% Diazinon Granules
Bioneem
Diazinon 2% Granules
Diazinon 5G
Diazinon 5% Granules
Diazinon 25%
Diazinon Granules
Diazinon Soil & Turf Insect Control
Dursban Ant & Turf
Dursban Lawn & Perimeter Granules
Flea & Tick Granules
Garden Insect Spray
Garden Soil Insecticide
Japanese Beetle Repellent
Malathion 50 Plus
Mole Cricket Killer
Organic Greenhouse House & Veg. Spray
Saf-T-Side
Sevin Granules
Sevin Liquid
Sevin Spray/Liquid
Thiodan Insect Spray
Worm Ender

SWISS CHARD

5% Diazinon Granules
Cutworm Earwig & Sowbug Bait
Cygon 2E
Diazinon 2% Granules
Diazinon 5% Granules
Diazinon 5G

USAGE OF PRODUCTS

USAGE OF PRODUCTS

INSECTICIDES

Diazinon Granules
Diazinon Soil & Turf Insect Control
Earwig Sowbug & Grasshopper Bait
Garden Soil Insecticide
Grasshopper Earwig & Sowbug Bait
Liquid Flowable Sevin
Liquid Sevin
Malathion 50 Plus
Mole Cricket Killer
Sevin 5% Dust
Sevin Granules
Sevin Liquid
Sevin Spray/Liquid
Tomato & Veg. Dust
Worm Ender

SYCAMORE

Diazinon Plus
Horticultural Spray Oil
Intercept H&G
Spray Oil

TARRAGON

Bioneem
Japanese Beetle Repellent

TAXUS (Yew)

Borer Miner Killer 5
Cygon 2E
Garden Insect Spray
Horticultural Spray Oil
Intercept H&G
Intercept Vegetable & Garden Spray
Permethrin T/O
Thiodan Insect Spray
Systemic Shrub & Flower Insecticide

TEA

Bug Buster-O

THUJA

Intercept H&G

THYME

Bioneem
Japanese Beetle Repellent

TOBACCO

Saf-T-Side
Sevin Dura Dust

TOBIRA

Intercept RFO

TOMATOES

2% Systemic Granules
5% Diazinon Granules
5% Malathion Dust
5% Sevin Garden Dust
25% Diazinon
25% Methoxychlor Spray
50% Malathion
Bacillus thuringiensis
Bioneem
Bio Worm Killer
Blue Dragon Garden Dust
Bug-B-Gon
Bug Buster
Caterpillar Killer
Colo. Potato Beetle Beater
Cutworm Earwig & Sowbug Bait
Cygon 2E
Diazinon 2% Granules
Diazinon 5G
Diazinon 5% G
Diazinon 5% Granules
Diazinon 12 1/2% E
Diazinon 25%
Diazinon 25% EC
Diazinon Dust
Diazinon Granules
Diazinon Plus
Diazinon Soil & Turf Insect Control
Diazinon Spray
Dipel
Dipel Dust
Earwig Sowbug & Grasshopper Bait
Flower Garden Insecticidal Soap
Fruit & Berry Insect Spray
Fruit & Veg Insect Spray
Garden Insect Spray
Garden Soil Insecticide
Garden Spray
Grasshopper Earwig & Sowbug Bait
House Plant & Garden Insect Spray
Insecticidal Soap Multi Purpose Insect Killer
Intercept Vegetable & Garden Spray
Japanese Beetle Killer
Japanese Beetle Repellent
Kelthane
Liquid Flowable Sevin
Liquid Sevin
Malathion 50%
Malathion 50 Plus
Many Purpose Insect Killer
Methoxychlor 25%
Mole Cricket Killer
Multi Purpose Insect Killer
Natural Pyrethrin
Organic Greenhouse House & Veg. Spray
Red Spider Spray
Roses & Flower Insect Spray
Saf-T-Side
Sevin 5 Dust
Sevin 10 Dust
Sevin 5% Dust
Sevin 10% Dust
Sevin Garden Dust
Sevin Granules
Sevin Liquid
Sevin 50 Wettable
Sevin 50% WP
Sevin 50 WP
Sevin Dura-Dust
Sevin Dura-Spray
Sevin Spray/Liquid
Sun Spray Ultra Fine
Systemic Granules
Thiodan Insect Spray
Thiodan Vegetable & Ornamental Dust
Thuricide
Tomato & Veg. Dust
Tomato & Veg. Fogger
Tomato & Vegetable Insect Killer
Tomato & Veg. RTU
Tomato Pepper Veg. Spray RTU
Vegetable Insect Control
Vegetable Insect Killer
Worm Ender
Yard & Garden Insect Killer

TREES (general)

25% Methoxychlor Insect Spray
50% Malathion
Bacillus thuringiensis
Bagworm & Mite Spray
Bioneem
Bio Worm Killer
Bug Buster-O
Caterpillar Killer
Diazinon 25% EC
Dipel
Dormant Spray
Dormant & Summer Oil Spray
Dursban
Dursban Insecticide
Dursban Insecticide Conc.
Dursban Insect Killer
Dursban Lawn Insect Killer
Fruit & Vegetable Insect Spray
Horticultural Spray Oil
Houseplant Insecticidal Soap
Houseplant & Garden Insect Spray
Imidan
Insecticidal Soap Mullti Purpose Insect Killer
Intercept Vegetable & Garden Spray
Isotox
Japanese Beetle Killer
Japanese Beetle Repellent
Kelthane
Lindane Borer Spray
Liquid Flowable Sevin

68

INSECTICIDES

USAGE OF PRODUCTS

Liquid Sevin
Malathion 50%
Many Purpose Dursban
  Concentrate
Methoxychlor 25%
Mite & Insect Spray
Orthene
Permethrin T/O
Pestkil
Rose & Flower Insect Spray
Rose & Flower Insect Killer II
Rose & Flower Insect Spray
Rose, Flower & Ornamental
  Insect Spray
Saf-T-Side
Scale Away
Sevin 5% Dust
Sevin 50 Wettable
Sevin Dura-Spray
Sevin Liquid
Spray Oil
Summer & Dormant Oil
Sun Spray Ultra Fine
Thiodan Insect Spray
Thuricide
Tomato & Vegetable Insect Killer
Volck Oil Spray
Whack Hornet & Wasp Killer
Worm Ender
Yard & Garden Insect Killer

TROPICAL FRUITS (general)
Bug Buster-O
Worm Ender

TROXON
Intercept RFO

TSUGA
Intercept H&G

TULIPS
Fruit & Veg Insect Spray
House Plant & Garden
  Insect Spray
Roses & Flower Insect Spray

TULIP POPLAR
Permethrin T/O

TULIPTREE
Horticultural Spray Oil

TURF GRASSES (general)
5% Diazinon Dust
5% Diazinon Granular
25% Diazinon
Ant Barrier
Ant Flea & Tick Insect Granules
Ant Flea & Tick Killer
Ant killer Granules
Ant Killer Powder

Bioneem
Bug Buster-O
Carpenter Ant Killer
Cutworm & Cricket Bait
Cutworm, Earwig & Sowbug Bait
Diazinon 2% Granules
Diazinon 4E
Diazinon 5% G
Diazinon 12 1/2% E
Diazinon 25%
Diazinon 25% EC
Diazinon 5G
Diazinon Insect Control
Diazinon Granules
Diazinon Plus
Diazinon Soil & Turf Insect Control
Diazinon Spray
Dursban
Dursban 1% Granules
Dursban Ant & Turf
Dursban Granules
Dursban Insecticide Conc.
Dursban Lawn Insect Control
Dursban Lawn Insect Killer
Dursban Lawn Insect Spray
Dursban Lawn & Perimeter
  Granules
Dursban Mole Cricket Bait
Dursban Plus
Dursban Ready Spray
Earwig Sowbug & Grasshopper
  Bait
Fire Ant Granules
Fire Ant Killer Granules
Fire Ant Killer Granules II
Flea-B-Gon Outdoor Flea & Tick
  Killer
Flea & Tick Granules
Home Pest Control Conc.
Intercept H&G
Intercept Vegetable & Granular
  Spray
Japanese Beetle Repellent
Kelthane
Lindane Borer Spray
Liquid Flowable Sevin
Liquid Sevin
Logic Fire Ant Bait
Many Purpose Dursban
  Concentrate
Many Purpose Insect Killer
Mite & Insect Spray
Oftanol
Orthene
Orthene Fire Ant Killer
Ortho-Klor
Permethrin T/O
Pronto
Sevin 5% Dust
Sevin 10% Dust
Sevin50% WP
Sevin 50 Wettable

Sevin Dura Spray
Sevin Granules
Sevin Liquid
Sevin Spray/Liquid
Termite Killer

TURNIPS
5% Diazinon Granules
5% Malathion Dust
25% Diazinon Insect Spray
50% Malathion
Ant, Flea & Tick Insect Granules
Bacillus thuringiensis
Bioneem
Bio Worm Killer
Blue Dragon Garden Dust
Bug-B-Gon
Cutworm Earwig & Sowbug Bait
Cygon 2E
Diazinon 2% Granules
Diazinon 5G
Diazinon 5% Granules
Diazinon 12 1/2% E
Diazinon 25%
Diazinon 25% EC
Diazinon Plus
Diazinon Soil & Turf Insect Control
Diazinon Spray
Dipel Dust
Dursban 1% Granules
Dursban Ant & Turf
Dursban Granules
Dursban Lawn & Perimeter
  Granules
Earwig Sowbug & Grasshopper
  Bait
Flea & Tick Granules
Fruit & Berry Insect Spray
Garden Soil Insecticide
Garden Spray
Grasshopper Earwig & Sowbug
  Bait
Japanese Beetle Killer
Japanese Beetle Repellent
Liquid Flowable Sevin
Liquid Sevin
Malathion 50%
Malathion 50 Plus
Mole Cricket Killer
Natural Pyrethrin
Sevin 5% Dust
Sevin 50 Wettable
Sevin Granules
Sevin Liquid
Sevin Spray/Liquid
Thiodan Vegetable & Ornamental
  Dust
Thuricide
Tomato & Veg. Fogger
Tomato & Veg Insect Killer
Tomato & Vegetable Insect Spray
Tomato & Veg. RTU

69

INSECTICIDES

Tomato Pepper Veg. Spray RTU
Vegetable Insect Control
Vegetable Insect Killer
Worm Ender

VANILLA
Aphid Mite & Whitefly Killer II
Spider Mite Control
Whitefly & Mealybug Control

VEGETABLES (general)
Bug Buster-O
Caterpillar Killer
Flower Garden Insecticidal Soap
Fruit & Vegetable Insect Killer
Houseplant Insecticidal Soap
Insecticidal Soap Multi Purpose Insect Killer
Insecticidal Soap for Fruit & Vegetables

VELVET PLANT
Intercept Vegetable & Garden Spray

VERBENA
Intercept H&G
Intercept RFP
Intercept Vegetable & Garden Spray
Sevin 5 Dust
Sevin 10 Dust
Spider Mite & Mealybug Control
Whitefly & Mealybug Spray

VINCA
Intercept H&G

VINES (general)
Dursban
Dursban Lawn Insect Killer
Fruit & Veg Insect Spray
House Plant & Garden Insect Spray
Roses & Flower Insect Spray

VIOLETS
5% Malathion Dust
Systemic Granules

VIRBURNUM
Bug-B-Gon
Intercept H&G
Intercept RFO
Multi Purpose Insect Killer
Sevin 5 Dust
Sevin 10 Dust
Soluable Oil Spray
Systemic Rose & Flower Food

WALNUTS
Bacillus thuringiensis
Bioneem
Diazinon 25%
Diazinon 25% EC
Diazinon Plus
Garden Insect Spray
Horticultural Spray Oil
Japanese Beetle RepellentMalathion 50%
Liquid Flowable Sevin
Red Spider Spray
Scale-Away
Sevin Dura Spray
Sevin Liquid
Sevin Spray/Liquid
Spray Oil
Summer & Dormant Oil
Worm Ender

WANDERING JEW
Aphid Mite & Whitefly Killer II
Fruit & Veg Insect Spray
Garden Spray
Gentle Care House Plant Spray
Home Patrol Insect Killer
House Plant & Garden Insect Spray
Intercept RFO
Japanese Beetle Killer
Mite Beater
Natural Pyrethrin
Plant Insect Control
Rose & Floral Insect Control
Rose & Floral Insect Killer
Rose & Flower Insect Spray
Rose & Flower Insect Killer II
Rose & Garden Insect Fogger
Rose Flower & Ornamental Insect Spray
Spider Mite Control
Spider Mite & Mealybug Control
Vegetable Insect Killer
Whitefly & Mealybug Control
Whitefly & Mealybug Spray

WATERCRESS
Worm Ender

WATERMELON
5% Diazinon Granules
25% Diazinon Insect Spray
Bacillus thuringiensis
Bioneem
Bug-B-Gon
Diazinon 2% Granules
Diazinon 5G
Diazinon 5% G
Diazinon 5% Granules
Diazinon 25%
Diazinon 25% EC

Diazinon Granules
Diazinon Plus
Garden Soil Insecticide
Insecticidal Soap Multi Purpose Insect Killer
Japanese Beetle Repellent
Kelthane
Malathion 50%
Mole Cricket Killer
Multi Purpose Insect Killer
Thuricide
Worm Ender

WAX PLANT
Aphid Mite & Whitefly Killer II
Intercept RFO
Intercept Vegetable & Garden Spray
Permethrin T/O
Spider Mite Control
Spider Mite & Mealybug Control
Whitefly & Mealybug Control
Whitefly & Mealybug Spray

WEEPING FIG
Aphid Mite & Whitefly Killer II
Intercept Vegetable & Garden Spray
Permethrin T/O
Spider Mite Control
Vegetable Insect Control
Whitefly & Mealybug Control

WIGELIA
Intercept H&G
Sevin 5 Dust
Sevin 10 Dust
Systemic Rose & Flower Food

WILLOW
25% Diazinon
Bagworm & Mite Spray
Borer Miner Killer 5
Diazinon 4E
Diazinon 12 1/2% E
Diazinon 25%
Diazinon 25% EC
Diazinon Plus
Diazinon Spray
Dormant Oil Spray
Horticultural Spray Oil
Imidan

WINTERGREEN
Bioneem
Japanese Beetle Repellent

WISTERIA
Rose & Flower Insect Killer II
Sevin 5 Dust
Sevin 10 Dust

INSECTICIDES

WOOD TREATMENT
Ortho-Klor

YAIPON
Intercept H&G

YEW
2% Systemic Granules
Imidan
intercept H&G
Intercept RFO
Sevin 5 Dust
Sevin 10 Dust
Systemic Granules
Systemic House Plant
  Insect Control

ZEBRA PLANT
Aphid Mite & Whitefly Killer II
Intercept RFO
Spider Mite Control
Spider Mite & Mealybug Control
Whitefly & Mealybug Control
Whitefly & Mealybug Spray

ZINNIAS
5% Diazinon Dust
5% Sevin Garden Dust
Ant Killer Powder
Fruits & Vegetable Insect Spray
Home Patrol Insect Killer
House Plant & Garden
  Insect Spray
Intercept H&G
Intercept RFO
Intercept Vegetable & Garden
  Spray
Liquid Sevin
Mite Beater
Permethrin T/O
Rose & Flower Dust
Roses & Flower Insect Spray
Sevin 5 Dust
Sevin 10 Dust
Sevin 50% WP
Sun Spray Ultra Fine
Systemic Granules
Systemic House Plant
  Insect Control
Whitefly & Mealybug Spray

USAGE OF PRODUCTS

HERBICIDES

ABBOTSWOOD
Grass Out

ABELIA
Amaze
Betasan 3.6
Garden Weed Preventer Granules
Surflan
Vegetable Turf & Ornamental
  Weeder
Weed Graules
Weed & Grass Preventer
Weed Stopper

ACACIA
Amaze
Grass Out
Poast
Surflan
Weed Stopper

ACALYPHA
Grass Out

AFRICAN DAISY
Poast
Weed Stopper

AGAPANTHUS
(Lily of the Nile)
Amaze
Grass Out
Poast
Surflan
Weed Stopper

AGERATUM
Eptam Weed Control
Eptam Weed & Grass Preventer
Garden Weed Preventer Granules
Grass-Out
Pre Emergent Weed & Grass
  Preventer
Vegetable & Turf Ornamental
  Weeder
Weed Granules

AGLANONEMA
Grass Out

AJUGA (Bugleweed)
Betasan 3.6
Eptam Weed Control
Eptam Weed & Grass Preventer
Grass Out
Grass-B-Gon
Poast
Pre Emergent Weed & Grass
  Preventer
Weed Stopper

ALMOND
Casoron
Grass-B-Gon
Poast
Surflan
Weed Stopper

ALOE
Grass Out

ALPINE CURRANT
Poast

ALYSSUM
Eptam Weed Control
Eptam Weed & Grass Preventer
Garden Weed Preventer Granules
Grass Out
Pre Emergent Weed & Grass
  Preventer
Poast
Vegetable Turf & Ornamental
  Weeder
Weed Granules

AMARANTHUS
Eptam Weed Control
Eptam Weed & Grass Preventer
Pre Emergent Weed & Grass
  Preventer

AMERICAN SWEETGUM
(Liquidambar)
Amaze
Grass-B-Gon
Poast
Surflan
Weed Stopper

AMUR CORKTREE
Casoron

APPLE
Casoron
Grass-B-Gon
Poast
Surflan
Weed Stopper

APRICOT
Grass-B-Gon
Surflan
Weed Stopper

ARBORVITAE
(American Thuja)
Amaze
Casoron
Garden Weed Preventer Granules
Grass-B-Gon
Poast
Surflan
Vegetable & Turf Ornamental
  Weeder
Weed & Grass Preventor
Weed Stopper

ARBOVITAE (Oriental)
Garden Weed Preventer Granules
Surflan
Weed Granules
Weed Stopper

ARIZONA CYPRESS
(Cupressus)
Poast
Weed Stopper

ARTICHOKE
Poast

ASH
Amaze
Casoron
Garden Weed Preventer Granules
Grass-B-Gon
Poast
Surflan
Vegetable & Turf Ornamental
  Weeder
Weed Granules
Weed Stopper

ASPARAGUS
Poast

ASTER
Amaze
Betasan 3.6
Epatm Weed Control
Eptam Weed & Grass Preventer
Garden Weed Preventer Granules
Pre Emergent Weed & Grass
  Preventer
Surflan
Vegetable Turf & Ornamental
  Weeder
Weed Granules

AUCUBA
(Gold Dust Plant)
Garden Weed Preventer Granules
Grass Out
Vegetable & Turf Ornamental
  Weeder
Weed Granules

AUREA
Grass Out

AUSTRALIA WILLOW
Grass Out

72

HERBICIDES

AVOCADO
Grass-B-Gon
Poast
Surflan
Weed Stopper

AZALEA
Amaze
Betasan 3.6
Casoron
Grass-B-Gon
Surflan
Weed & Grass Preventor
Weed Granules
Weed Stopper

AZARA
Betasan 3.6

BABY'S BREATH
Amaze
Garden Weed Preventer Granules
Surflan
Vegetable & Turf Ornamental
   Weeder
Weed Granules
Weed Stopper

BACHELOR'S BUTTON
Betasan 3.6

BALSAM
Eptam Weed Control
Eptam Weed & Grass Preventer
Pre Emergent Weed & Grass
   Preventer

BAMBOO
Amaze
Grass Out

BANANA
Grass Out

BANKSIA
Grass Out

BARBERRY (Berberia)
Amaze
Casoron
Eptam Weed & Grass Preventer
Eptam Weed Control
Garden Weed Preventer Granules
Grass-B-Gon
Poast
Pre Emergent Weed & Grass
   Preventer
Surflan
Vegetable & Turf Ornamental
   Weeder
Weed Granules
Weed Stopper

BASSWOOD
Poast

BEANS (Snap/Pole/Bush)
Eptam Weed Control
Garden Turf & Ornamental
   Herbicide
Garden Weed Preventer Granules
Poast
Vegetable & Turf Ornamental
   Weeder
Weed Granules

BEANS (Mung)
Vegetable & Turf Ornamental
   Weeder

BEANS (Blackeyes)
Vegetable & Turf Ornamental
   Weeder

BEARBERRY
Grass Out

BEAUTY BUSH
Casoron

BEECH
Grass-B-Gon

BEGONIA
Eptam Weed Control
Eptam Weed & Grass Preventer
Poast
Pre Emergent Weed & Grass
   Preventer

BELLFLOWER (Campanula)
Amaze
Betasan 3.6
Garden Weed Preventer Granules
Grass Out
Surflan
Vegetable & Turf Ornamental
   Weeder
Weed Granules
Weed Stopper

BERKMANS (Thuja)
Grass Out
Poast

BIRCH (Betula)
Amaze
Casoron
Garden Weed Preventer Granules
Grass-B-Gon
Grass Out
Poast
Surflan
Vegetable & Turf Ornamental
   Weeder
Weed Granules
Weed Stopper

BIRD OF PARADISE
Amaze
Grass Out
Poast
Surflan
Weed & Grass Preventer
Weed Stopper

BITTERSWEET
Poast

BLACK-EYED SUSAN (Rudbeckia)
Amaze
Surflan
Weed Stopper

BLAZING STAR (Liatris)
Amaze
Surflan
Weed Stopper

BLEEDING HEART
Amaze
Garden Weed Preventer Granules
Grass Out
Poast
Surflan
Vegetable & Turf Ornamental
   Weeder
Weed Granules
Weed Stopper

BLOODLEAF
Garden Weed Preventer Granules
Vegetable & Turf Ornamental
   Weeder
Weed Granules

BLUEBERRY
Casoron
Grass Out
Poast
Surflan
Weed Stopper

BLUE FESCUE
Amaze

BLUE HIBISCUS (Aloyogrne)
Poast

BLUE STAR CREEPER
Grass Out

USAGE OF PRODUCTS

73

HERBICIDES

BOTTLE TREE
Grass Out
Poast

BOTTLEBRUSH
Amaze
Grass Out
Poast
Surflan
Weed Stopper

BOUGAINVILLEA
Amaze
Grass Out
Weed Stopper

BOX ELDER
Casoron

BOXWOOD (Buxus)
Amaze
Betasan 3.6
Casoron
Eptam Weed Control
Eptam Weed & Grass Preventer
Garden Weed Preventer Granules
Grass-B-Gon
Grass Out
Poast
Pre Emergent Weed & Grass Preventer
Surflan
Vegetable & Turf Ornamental Weeder
Weed & Grass Preventor
Weed Granules
Weed Stopper

BRAZILIAN PEPPER TREE
Grass Out
Poast

BRISBANE BOX TREE
Poast

BROCCOLI
Garden Turf & Ornamental Herbicide
Garden Weed Preventer Granules
Poast
Vegetable Turf & Ornamental Weeder
Weed Granules

BROOM (Cytisus)
Amaze
Grass Out
Weed Stopper

BRUSH CHERRY (Eugenia)
Grass Out
Surflan
Weed & Grass Preventer
Weed Stopper

BRUSH CONTROL
Brush Buster
Roundup

BRUSSELS SPROUTS
Garden Turf & Ornamental Herbicide
Garden Weed Preventer Granules
Weed Granules

BUGLOSS
Garden Weed Preventer Granules
Vegetable & Turf Ornamental Weeder
Weed Granules

BULBS (general)
Amaze

BURNINGBUSH
Amaze
Grass Out

BUSH CHERRY
Amaze
Grass Out

BUSH DAISY
Poast

CABBAGE
Garden Turf & Ornamental Herbicide
Garden Weed Preventer Granules
Poast
Vegetable Turf & Ornamental Weeder
Weed Granules

CACTUS
Amaze
Grass Out
Poast
Surflan

CAJEPUT TREE
Poast

CALADIUM
Amaze
Surflan
Weed Stopper

CALENDULA
Betasan 3.6

CALIFORNIA PEPPER TREE
Amaze
Grass Out

CAMELLIA
Casoron
Eptam Weed Contol
Eptam Weed & Grass Preventer
Garden Weed Preventor Granules
Grass Out
Poast
Pre Emergent Weed & Grass Preventer
Vegetable Turf & Ornamental Weeder
Weed Granules

CANDYTUFT
Betasan 3.6
Garden Weed Preventer Granules
Grass Out
Poast
Vegetable Turf & Ornamental Weeder
Weed Granules

CANEBERRIES
Casoron
Grass-B-Gon
Poast
Surflan
Weed Stopper

CAPE MARIGOLD
Surflan
Weed Stopper

CAPEWEED (Arctotheca)
Amaze
Grass Out
Weed Stopper

CARAGANA
Casoron

CARICATURE PLANT
Grass Out

CAROB
Amaze
Grass Out
Poast
Weed Stopper

CAROLINE CHERRY
Grass Out

CARPET BUGLE
Amaze
Surflan

HERBICIDES

USAGE OF PRODUCTS

CARROTWOOD
- Amaze
- Grass Out
- Poast
- Weed Stopper

CASSIA
- Grass Out
- Poast
- Surflan
- Weed Stopper

CATALPA
- Poast

CAULIFLOWER
- Garden Turf & Ornamental Herbicide
- Garden Weed Preventer Granules
- Poast
- Vegetable Turf & Ornamental Weeder
- Weed Granules

CEANOTHUS (Jersey Tea/Wild Lilac)
- Grass Out
- Poast
- Surflan
- Weed Stopper

CEDAR
- Amaze
- Casoron
- Grass Out
- Weed & Grass Preventor

CELERY
- Poast

CENTURY PLANT (Agave)
- Amaze
- Grass Out
- Surflan
- Weed Stopper

CHAMAECYPARIS (False Cypress)
- Eptam Weed Control
- Eptam Weed & Grass Preventer
- Surflan
- Weed Stopper

CHERRIES
- Casoron
- Grass-B-Gon
- Poast
- Surflan
- Weed Stopper

CHERRY (Carolina)
- Poast

CHERRY (Flowering)
- Amaze
- Grass-B-Gon
- Poast
- Weed Stopper

CHESTNUT
- Garden Weed Preventer Granules
- Vegetable Turf & Ornamental Weeder
- Weed Granules

CHIVES
- Grass Out

CHOKECHERRY
- Poast

CHRISTMAS TREES
- Surflan
- Weed Stopper

CHRYSANTHEMUM
- Amaze
- Eptam Weed Control
- Eptam Weed & Grass Preventer
- Garden Weed Preventer Granules
- Poast
- Pre Emergent Weed & Grass Preventer
- Surflan
- Vegetable Turf & Ornamental Weeder
- Weed Granules
- Weed & Grass Preventer
- Weed Stopper

CITRUS
- Eptam Weed Control
- Eptam Weed & Grass Preventer
- Grass Out
- Poast
- Pre Emergent Weed & Grass Preventer
- Surflan
- Weed Stopper

CLEYERA
- Amaze
- Casoron
- Grass Out
- Surflan
- Weed Stopper

COCKSCOMB (Celosia)
- Poast

COLEUS
- Garden Weed Preventer Granules
- Grass-B-Gon
- Grass Out
- Poast
- Vegetable Turf & Ornamental Weeder
- Weed Granules

COLLARDS
- Garden Turf & Ornamental Herbicide
- Garden Weed Preventer Granules
- Vegetable Turf & Ornamental Weeder
- Weed Granules

COLUMBINE
- Garden Weed Preventer Granules
- Grass Out
- Vegetable Turf & Ornamental Weeder
- Weed Granules

CONEFLOWER
- Amaze
- Garden Weed Preventer Granules
- Surflan
- Vegetable Turf & Ornamental Weeder
- Weed Granules
- Weed Stopper

CORAL BEAUTY (Cotoneaster)
- Grass Out
- Poast

CORAL BELLS
- Weed Granules

CORDYLINE
- Grass Out

COREOPSIS
- Amaze
- Garden Weed Preventer Granules
- Grass Out
- Surflan
- Vegetable Turf & Ornamental Weeder
- Weed Granules
- Weed Stopper

CORNUS
- Grass Out

COSMOS
- Weed Granules

75

HERBICIDES

COTONEASTER
Amaze
Casoron
Garden Weed Preventer Granules
Grass-B-Gon
Grass Out
Poast
Surflan
Vegetable Turf & Ornamental Weeder
Weed Granules
Weed & Grass Preventer
Weed Stopper

COTTONWOOD
Amaze
Garden Weed Preventer Granules
Surflan
Vegetable Turf & Ornamental Weeder
Weed Granules
Weed Stopper

COYOTEBUSH (Baccharis)
Amaze
Surflan
Weed Stopper

CRABAPPLE
Amaze
Casoron
Garden Weed Preventer Granules
Grass-B-Gon
Grass Out
Poast
Surflan
Vegetable Turf & Ornamental Weeder
Weed Granules
Weed Stopper

CRANBERRY BUSH (Viburnum)
Amaze
Garden Weed Preventer Granules
Poast
Weed Stopper

CRAPE MYRTLE
Amaze
Poast
Surflan
Weed Stopper

CRIMSON PYGMY (Berberis)
Grass Out
Poast

CROSSANDRA
Grass Out

CROTON
Grass out

CROWNVETCH
Grass Out
Poast

CRYPTOMERIA (Japanese)
Amaze
Surflan
Weed Stopper

CUCUMBERS
Garden Turf & Ornamental Herbicide
Garden Weed Preventer Granules
Poast
Vegetable Turf & Ornamental Weeder

CUPHEA (False Heather)
Garden Weed Preventer Granules
Poast
Vegetable Turf & Ornamental Weeder
Weed Granules

CURRANT
Weed Stopper

CYPRESS (Chamaecyparis)
Amaze
Betasan 3.6
Casoron
Garden Weed Preventer Granules
Grass Out
Poast
Pre Emergent Weed & Grass Preventer
Surflan
Vegetable Turf & Ornamental Weeder
Weed Granules
Weed Stopper

DAFFODIL
Betasan 3.6
Surflan
Weed & Grass Preventer
Weed Stopper

DAHLIA
Betasan 3.6
Eptam Weed Control
Eptam Weed & Grass Preventer
Garden Weed Preventer Granules
Poast
Pre Emergent Weed & Grass Preventer
Vegetable Turf & Ornamental Weeder
Weed Granules

DAISY
Amaze
Betasan 3.6
Surflan
Weed & Grass Preventer
Weed Stopper

DAISY (Shasta)
Amaze
Poast
Surflan
Weed Stopper

DATES
Grass-B-Gon

DAYLILY
Amaze
Eptam Weed Control
Eptam Weed & Grass Preventer
Poast
Pre Emergent Weed & Grass Preventer
Surflan
Weed Stopper

DELPHINIUM
Garden Weed Preventer Granules
Vegetable Turf & Ornamental Weeder
Weed Granules

DESERT FERN TREE
Grass Out

DEUTZIA
Amaze
Casoron
Garden Weed Preventer Granules
Grass Out
Vegetable Turf & Ornamental Weeder
Weed Granules
Weed Stopper

DEWBERRY
Weed Stopper

DIANTHUS
Eptam Weed Control
Eptam Weed & Grass Peventer
Poast

DICHRONDA
Crabgrass Killer
Grass-B-Gon

HERBICIDES

USAGE OF PRODUCTS

DOGWOOD
Amaze
Casoron
Eptam Weed Control
Eptam Weed & Grass Preventer
Garden Weed Preventer Granules
Grass-B-Gon
Grass Out
Poast
Pre Emergent Weed & Grass Preventer
Surflan
Vegetable Turf & Ornamental Weeder
Weed Granules
Weed Stopper

DRACAERA
Grass Out

DUMBCANE
Grass Out

DUSTY MILLER
Grass Out
Poast

EASTER CACTUS
Weed Stopper

EGGPLANT
Garden Turf & Ornamental Herbicide
Garden Weed Preventer Granules
Poast
Vegetable Turf & Ornamental Weeder

ELDERBERRY
Surflan
Weed Stopper

ELM
Amaze
Casoron
Garden Weed Preventer Granules
Grass-B-Gon
Poast
Vegetable Turf & Ornamental Weeder
Weed Granules
Weed Stopper

ESCALLONIA
Amaze
Grass Out
Surflan
Weed Stopper

EUCALYPTUS (Gumtree)
Amaze
Garden Weed Preventer Granules
Grass-B-Gon
Grass Out
Poast
Surflan
Vegetable Turf & Ornamental Weeder
Weed Granules
Weed & Grass Preventer
Weed Stopper

EUONYMUS
Amaze
Casoron
Eptam Weed Control
Eptam Weed & Grass Preventer
Garden Weed Preventer Granules
Grass-B-Gon
Grass Out
Poast
Pre Emergent Weed & Grass Preventer
Surflan
Vegetable Turf & Ornamental Weeder
Weed Granules
Weed & Grass Preventor
Weed Stopper

EUPHORBIA
Grass Out

EVENING PRIMROSE
Vegetable Turf & Ornamental Weeder

FALSE SPIRAEA (Astilbe)
Grass Out
Surflan
Weed Stopper

FATSHEDERA
Amaze
Grass Out
Surflan
Weed & Grass Preventer
Weed Stopper

FERN (Asparagus)
Poast

FESCUE (Blue)
Weed Stopper

FETTERBUSH
Grass Out

FEVERFEW
Garden Weed Preventer Granules
Vegetable Turf & Ornamental Weeder
Weed Granules

FICUS (Weeping Fig)
Amaze
Grass Out
Poast
Weed Stopper

FIDDLEWOOD
Grass Out

FIG
Grass-B-Gon
Poast
Surflan
Weed Stopper

FILBERT
Casoron
Grass-B-Gon
Surflan
Weed Stopper

FIR
Amaze
Eptam Weed Control
Eptam Weed & Grass Preventer
Garden Weed Preventer Granules
Grass-B-Gon
Grass Out
Poast
Pre Emergent Weed & Grass Preventer
Surflan
Vegetable Turf & Ornamental Weeder
Weed Granules
Weed & Grass Preventer
Weed Stopper

FIR (Douglas)
Grass-B-Gon
Grass Out
Poast
Weed & Grass Preventer

FIREWHEEL TREE
Grass Out

FLOWERING MAPLE (Abutilon)
Weed Stopper

FLOWERS (general)
Amaze
Garden Turf & Ornamental Herb.
Hose N Go Weed & Grass Preventer
Weed Granules

FORGET-ME-NOT
Garden Weed Preventer Granules

77

HERBICIDES

Vegetable Turf & Ornamental
  Weeder
Weed Granules

FORSYTHIA
Amaze
Casoron
Garden Weed Preventer Granules
Grass Out
Poast
Surflan
Vegetable Turf & Ornamental
  Weeder
Weed Granules
Weed & Grass Preventer
Weed Stopper

FOUNTAIN GRASS
Grass Out

FOUR O'CLOCK
Garden Weed Preventer Granules
Weed Granules

FOXGLOVE
Garden Weed Preventer Granules
Vegetable Turf & Ornamental
  Weeder
Weed Granules

FREESIA
Betasan 3.6

FUCHSIA
Poast

GARDENIA
Amaze
Casoron
Grass-B-Gon
Grass Out
Poast
Surflan
Weed & Grass Preventer
Weed Stopper

GAILLARDIA
Vegetable Turf & Ornamental
  Weeder
Weed Granules

GARLIC
Garden Turf & Ornamental
  Herbicide
Garden Weed Preventer Granules
Poast
Vegetable Turf & Ornamental
  Weeder

GARZANIA
Amaze
Betasan 3.6
Eptam Weed Control
Eptam Weed & Grass Preventer
Grass-B-Gon
Grass Out
Poast
Pre Emergent Weed & Grass
  Preventer
Surflan
Weed & Grass Preventer
Weed Stopper

GAY FEATHER
Grass Out

GERANIUM
Amaze
Garden Weed Preventer Granules
Grass-B-Gon
Grass Out
Poast
Surflan
Vegetable Turf & Ornamental
  Weeder
Weed Granules
Weed & Grass Preventer
Weed Stopper

GERBERA
Poast

GEUM
Amaze
Surflan
Weed Stopper

GINKGO (Maidenhair Tree)
Amaze
Surflan
Weed & Grass Preventer
Weed Stopper

GLADIOLUS
Amaze
Betasan 3.6
Garden Weed Preventer Granules
Grass Out
Poast
Surflan
Vegetable Turf & Ornamental
  Weeder
Weed Granules
Weed & Grass Preventer
Weed Stopper

GOLD DROP
Grass Out

GOLDENRAIN TREE
Amaze
Casoron
Grass-B-Gon
Surflan
Weed & Grass Preventer
Weed Stopper

GOLDENTUFT
Vegetable Turf & Ornamental
  Weeder
Weed Granules

GOOSEBERRY
Surflan
Weed Stopper

GRAPE
Casoron
Grass-B-Gon
Poast
Surflan
Weed Stopper

GRAPE IVY (Cissus)
Poast

GREEN CARPET
Grass Out

GROUND COVERS (general)
Amaze

GUAVA
Grass Out
Poast

GUINEA GOLD VINE (Hibberta)
Poast

HACKBERRY
Casoron
Poast

HAREBELL
Poast

HAWTHORN
Amaze
Garden Weed Preventer Granules
Vegetable Turf & Ornamental
  Weeder
Weed Granules

HEARTS & FLOWERS
Grass Out

HEATH
Garden Weed Preventer Granules
Vegetable Turf & Ornamental
  Weeder
Weed Granules

HEATHER
Casoron
Grass Out

HERBICIDES

USAGE OF PRODUCTS

HEMLOCK
Eptam Weed Control
Eptam Weed & Grass Preventer
Grass-B-Gon
Grass Out
Poast
Pre Emergent Weed & Grass Preventer

HEN & CHICKENS
Grass Out

HIBISCUS (Rose of Sharon)
Amaze
Grass Out
Poast
Surflan
Weed Stopper

HOLLY
Amaze
Casoron
Eptam Weed Control
Eptam Weed & Grass Preventer
Garden Weed Preventer Granules
Grass-B-Gon
Grass Out
Poast
Pre Emergent Weed & Grass Preventer
Surflan
Vegetable Turf & Ornamental Weeder
Weed Graules
Weed & Grass Preventor
Weed Stopper

HOLLY (Yaupon)
Poast
Weed Stopper

HOLLYHOCK
Grass Out

HONEYLOCUST
Grass Out
Poast
Weed Stopper

HONEYSUCKLE
Amaze
Casoron
Garden Weed Preventer Granules
Grass Out
Poast
Surflan
Vegetable Turf & Ornamental Weeder
Weed Granules
Weed & Grass Preventor
Weed Stopper

HONEYSUCKLE (Bush) (Dierville)
Poast

HONEYSUCKLE (Cape) (Tecomaria)
Poast

HONEYSUCKLE (Mexican/Justicia)
Garden Weed Preventer Granules
Grass Out
Surflan
Weed Stopper

HONEYSUCKLE (Trumpet)
Garden Weed Preventer Granules
Weed Stopper

HONEYSUCKLE (Winter)
Garden Weed Preventer Granules
Weed Stopper

HOPSEEDBUSH (Dodonaea)
Amaze
Grass Out
Poast
Surflan
Weed Stopper

HORSE CHESTNUT
Grass-B-Gon

HORSERADISH
Garden Turf & Ornamental Herbicide
Garden Weed Preventer Granules
Vegetable Turf & Ornamental Weeder

HOSTA
Grass Out
Poast
Weed Stopper

HYACINTH
Surflan
Weed & Grass Preventer
Weed Stopper

HYDRANGEA
Garden Weed Preventer Granules
Grass Out
Poast
Vegetable Turf & Ornamental Weeder
Weed Granules

HYPERICUM (St. Johns Wort)
Amaze
Betasan 3.6
Eptam Weed Control
Eptam Weed & Grass Preventer
Grass-B-Gon
Pre Emergent Weed & Grass Preventer
Weed Stopper

ICE PLANT
Amaze
Betasan 3.6
Eptam Weed Control
Eptam Weed & Grass Preventer
Pre Emergent Weed & Grass Preventer
Grass Out
Surflan
Weed & Grass Preventer

ICEPLANT (Carpobrotus)
Surflan
Weed Stopper

ICEPLANT (Trailing)
Weed Stopper

ICEPLANT (White)
Weed Stopper

IMPATIENS
Amaze
Poast
Surflan
Weed & Grass Preventer
Weed Stopper

INDIAN LAUREL (Ficus)
Grass Out
Poast

INKBERRY
Grass Out

IRIS
Amaze
Garden Weed Preventer Granules
Grass-B-Gon
Grass Out
Poast
Surflan
Weed Granules
Weed & Grass Preventer
Weed Stopper

IRONBARK
Poast

79

HERBICIDES

IRONWOOD
Grass Out

IVY
Amaze
Betasan 3.6
Casoron
Eptam Weed Control
Eptam Weed & Grass Preventer
Grass-B-Gon
Grass Out
Pre Emergent Weed & Grass
  Preventer
Surflan
Weed Granules
Weed & Grass Preventor
Weed Stopper

IVY (Boston)
Poast
Weed Granules

IVY (Hedera)
Garden Weed Preventer Granules
Poast
Surflan
Weed Stopper

IXORA
Grass Out

JACARANDA
Grass Out
Poast

JACK IN THE PULPIT
Poast

JACKMANNI
Grass Out

JADE PLANT
Grass Out
Poast

JASMINE (Orange)
Amaze
Grass Out
Poast

JESSAMINE (Carolina)
Amaze
Grass Out
Poast

JOJOBA
Grass Out
Poast

JUNIPER
Amaze
Betasan 3.6
Casoron
Eptam Weed Control
Eptam Weed & Grass Preventer
Garden Weed Preventer Granules
Grass-B-Gon
Grass Out
Poast
Pre Emergent Weed & Grass
  Preventer
Surflan
Vegetable Turf & Ornamental
  Weeder
Weed Granules
Weed & Grass Preventor
Weed Stopper

KALANCHOE
Grass Out

KALE
Garden Turf & Ornamental
  Herbicide
Garden Weed Preventer Granules
Vegetable Turf & Ornamental
  Weeder
Weed Granules

KINNIKINNICH
Casoron
Grass-B-Gon

KIWI
Surflan
Weed Stopper

KNOTWEED
Grass Out

KUMQUAT
Surflan

LANTANA
Garden Weed Preventer Granules
Grass Out
Poast
Vegetable Turf & Ornamental
  Weeder
Weed Granules

LARCH
Poast

LARKSPUR
Garden Weed Preventer Granules
Vegetable Turf & Ornamental
  Weeder

LAUCOTHOE
Weed Stopper

LAUREL
Amaze
Casoron
Weed & Grass Preventor

LAUREL (California)
Surflan
Weed Stopper

LAUREL CHERRY
Amaze
Surflan
Weed Stopper

LAVENDER
Poast

LAVENDER COTTON (Santolina)
Garden Weed Preventer Granules
Grass Out
Poast
Weed Granules

LAVENDER SCALLOPS
Grass Out

LAWN RENOVATION
Liquid Edger
Weed Ender

LENTILS
Poast

LETTUCE
Poast

LEUCOTHOE
Casoron
Eptam Weed Control
Eptam Weed & Grass Preventer
Pre Emergent Weed & Grass
  Preventer
Surflan
Weed Stopper

LILAC (Syringa)
Amaze
Casoron
Eptam Weed Control
Eptam Weed & Grass Preventer
Garden Weed Preventer Granules
Grass Out
Poast
Pre Emergent Weed & Grass
  Preventer
Surflan
Weed Granules
Weed & Grass Preventor
Weed Stopper

HERBICIDES

LILY (General)
Garden Weed Preventer Granules
Grass Out
Surflan
Vegetable Turf & Ornamental Weeder
Weed Granules

LILY OF THE VALLEY
Amaze
Grass Out
Poast

LILY OF THE VALLEY (Pieris)
Weed Stopper

LILY TURF (Liriope)
Amaze
Grass-B-Gon
Grass-Out
Poast
Surflan
Weed Stopper

LINDEN (Telia)
Amaze
Casoron
Eptam Weed Control
Eptam Weed & Grass Preventer
Poast
Pre Emergent Weed & Grass Preventer
Surflan
Weed Stopper

LOBELIA
Poast

LOCUST
Amaze
Casoron
Garden Weed Preventer Granules
Vegetable Turf & Ornamental Weeder
Weed Granules

LOQUAT
Poast

LUPINE
Garden Weed Preventer Granules
Vegetable Turf & Ornamental Weeder
Weed Granules

MACADAMIA NUT
Grass-B-Gon
Surflan
Weed Stopper

MAGNOLIA
Amaze
Casoron
Eptam Weed Control
Eptam Weed & Grass Preventer
Garden Weed Preventer Granules
Grass-B-Gon
Grass Out
Pre Emergent Weed & Grass Preventer
Poast
Surflan
Vegetable Turf & Ornamental Weeder
Weed Granules
Weed Stopper

MAHOGANY
Amaze
Weed Stopper

MANZANITA
Amaze
Surflan
Weed & Grass Preventer
Weed Stopper

MAPLE (Acer)
Amaze
Casoron
Eptam Weed Control
Eptam Weed & Grass Preventer
Garden Weed Preventer Granules
Grass-B-Gon
Grass Out
Poast
Pre Emergent Weed & Grass Preventer
Surflan
Vegetable Turf & Ornamental Weeder
Weed Granules
Weed & Grass Preventer
Weed Stopper

MARGUERITE
Garden Weed Preventer Granules
Weed Granules

MARIGOLD
Amaze
Betasan 3.6
Eptam Weed Control
Eptam Weed & Grass Preventer
Garden Weed Preventer Granules
Grass-B-Gon
Grass Out
Poast
Pre Emergent Weed & Grass Preventer
Surflan

Vegetable Turf & Ornamental Weeder
Weed Granules
Weed & Grass Preventer
Weed Stopper

MELONS
Garden Turf & Ornamental Herbicide
Garden Weed Preventer Granules
Poast
Vegetable Turf & Ornamental Weeder

MESQUITE
Grass Out

MEXICAN INDIGO
Grass Out

MICKEY MOUSE BUSH
Poast

MIMOSA TREE (Sensitive Plant)
Poast

MIRROR PLANT
Grass Out
Poast

MOCKORANGE
Amaze
Casoron
Garden Weed Preventer Granules
Poast
Surflan
Vegetable Turf & Ornamental Weeder
Weed Granules
Weed Stopper

MONDO GRASS
Amaze
Poast
Weed Stopper

MONEYWORT
Grass Out
Poast

MOON GLOW
Grass Out

MORNING GLORY
Garden Weed Preventer Granules
Grass Out
Vegetable Turf & Ornamental Weeder
Weed Granules

USAGE OF PRODUCTS

81

HERBICIDES

MOSS ROSE (Portulaca)
Amaze
Garden Weed Preventer Granules
Poast
Surflan
Vegetable Turf & Ornamental
    Weeder
Weed Granules
Weed Stopper

MOTHER OF THYME
Garden Weed Preventer Granules
Vegetable Turf & Ornamental
    Weeder
Weed Granules

MOUNTAIN ASH
Casoron
Grass-B-Gon
Poast

MOUNTAIN LAUREL
Amaze
Garden Weed Preventer Granules
Surflan
Vegetable Turf & Ornamental
    Weeder
Weed Granules
Weed Stopper

MOURNING BRIDE
Garden Weed Preventer Granules
Vegetable Turf & Ornamental
    Weeder
Weed Granules

MULBERRY (Morus)
Amaze
Weed Stopper

MUNGBEANS
Garden Weed Preventer Granules

MUSTARD GREENS
Garden Turf & Ornamental
    Herbicide
Garden Weed Preventer Granules
Vegetable Turf & Ornamental
    Weeder
Weed Granules

MYOPORUM
Amaze
Grass Out
Poast
Surflan
Weed Stopper

MYRTLE
Amaze
Betasan 3.6
Poast
Surflan
Weed Stopper

NANDINA (Heavenly Bamboo)
Amaze
Casoron
Grass Out
Poast
Surflan
Weed & Grass Preventer
Weed Stopper

NANNYBERRY (Viburnum)
Poast

NARCISSUS
Betasan 3.6
Surflan
Weed & Grass Preventer
Weed Stopper

NASTURTIUM
Eptam Weed Control
Eptam Weed & Grass Preventer
Garden Weed Preventer Granules
Pre Emergent Weed & Grass
    Preventer
Vegetable Turf & Ornamental
    Weeder
Weed Granules

NECTARINE
Casoron
Grass-B-Gon
Poast
Surflan
Weed Stopper

NEW ZEALAND CHRISTMAS TREE
Grass Out

NEW ZEALAND FLAX
Poast

NINEBARK (Physocarpus)
Grass Out
Poast

NON-CROP AREAS
2,4-D Amine
Blackberry & Brush Killer
Brush-B-Gon
Brush-B-Gon RTU
Brush Buster
Brushkil
Brush Killer
Brush-No-More
Com-Pleet
Crask & Crevise Weed Killer
Finale Concentrate
Finale Ready to Use
Finale Super Conc.
Granular Noxall Veg. Killer
Grass & Weed Killer
Grass & Weed Killer Kleen Up
    RTU
Grass & Weed Killer RTU
Grass Weed & Vegetable Killer
Ground Clear Triox
Ground Clear Super Edger
Home Deck & Patio Moss & Algae
    Killer
Kleenaway
Kleenaway Grass & Weed Killer
    RTU
Kleenup Grass & Weed Killer
Kleenup Weed & Grass Killer
Lawn Weed Killer
Liquid Edger
Moss Kil
Moss Kil Granules
Noxall Vegetation Killer
Poison Ivy & Brush Killer
Poison Ivy Bison Oak Killer
Poison Oak & Ivy Killer RTU
Roundup
RTU Knock-Out Weed & Grass
    Killer
RTU Moss Kil
Super Fast Weed & Grass Killer
Surflan
Total Vegetation Killer
Vegetation Killer
Weed & Grass Killer
Weed & Grass Killer Concentrate
Weed & Grass Killer RTU
Weed Ender
Weed Stopper
Weed Warrior
Weed Whacker
Weed Whacker Jet Spray
Yard Basics Weed & Grass Killer

OAK
Amaze
Casoron
Eptam Weed Control
Eptam Weed & Grass Preventer
Garden Weed Preventer Granules
Grass-B-Gon
Grass Out
Poast
Pre Emergent Weed & Grass
    Preventer
Surflan
Weed Granules
Weed & Grass Preventer
Weed Stopper

OCOTILLO
Grass Out

HERBICIDES

USAGE OF PRODUCTS

OLEANDER
- Amaze
- Grass Out
- Poast
- Surflan
- Weed & Grass Preventer
- Weed Stopper

OLIVE
- Amaze
- Grass-B-Gon
- Grass Out
- Poast
- Surflan
- Weed Stopper

ONIONS
- Garden Turf & Ornamental Herbicide
- Garden Weed Preventer Granules
- Poast
- Vegetable Turf & Ornamental Weeder

OREGON GRAPE (Mahonia)
- Amaze
- Grass Out
- Surflan
- Weed & Grass Preventer
- Weed Stopper

ORNAMENTALS (General)
- Amaze
- Finale Concentrate
- Finale RTU
- Finale Super Concentrate
- Garden Turf & Ornamental Herbicide
- Turf & Ornamental Herb.
- Weed Ender

ORNAMENTALS (before planting)
- Klearaway
- Roundup
- Weed Ender

ORPINE
- Garden Weed Preventer Granules
- Vegetable Turf & Ornamental Weeder
- Weed Granules

OSAGE ORANGE
- Poast

OSMANTHUS (Holly Olive)
- Amaze
- Casoron
- Grass Out
- Poast

- Surflan
- Weed Stopper

OYSTER PLANT
- Grass Out

PACHISTIMA
- Casoron
- Garden Weed Preventer Granules
- Vegetable Turf & Ornamental Weeder

PACHYSANDRA (Japanese Spurge)
- Betasan 3.6
- Eptam Weed Control
- Eptam Weed & Grass Preventer
- Garden Weed Preventer Granules
- Grass-B-Gon
- Grass Out
- Poast
- Pre Emergent Weed & Grass Preventer
- Vegetable Turf & Ornamental Weeder
- Weed Granules
- Weed Stopper

PAGODA FLOWER
- Grass Out

PALM
- Amaze
- Grass Out
- Weed Stopper

PALM (Areca)
- Weed Stopper

PALM (Chinese)
- Weed Stopper

PALM (Christmas)
- Weed Stopper

PALM (Date)
- Poast
- Weed Stopper

PALM (Fan)
- Poast

PALM (Queen)
- Poast
- Weed Stopper

PALM (Windmill)
- Poast
- Weed Stopper

PALO VERDE
- Amaze

- Grass Out
- Poast
- Surflan
- Weed Stopper

PAMPAS GRASS
- Amaze
- Grass Out
- Weed Stopper

PANSY (Viola)
- Amaze
- Betasan 3.6
- Eptam Weed Control
- Eptam Weed & Grass Preventer
- Poast
- Pre Emergent Weed & Grass Preventer
- Surflan
- Weed & Grass Preventer
- Weed Stopper

PASSION VINE
- Grass Out

PAULOWNIA
- Poast

PAXISTIONA
- Weed Granules

PEACH
- Casoron
- Grass-B-Gon
- Poast
- Surflan
- Weed Stopper

PEACH (Flowering)
- Grass-B-Gon

PEANUTS
- Poast

PEAR
- Casoron
- Grass-B-Gon
- Poast
- Surflan
- Weed Stopper

PEAR (Ornamental)
- Grass-B-Gon
- Grass Out

PEAS
- Poast

PECAN
- Casoron
- Grass-B-Gon
- Poast

83

HERBICIDES

Surflan
Weed Stopper

PEONY
Garden Weed Preventer Granules
Vegetable Turf & Ornamental
  Weeder
Weed Granules

PEPEROMIA
Grass Out

PEPPERS
Garden Turf & Ornamental
  Herbicide
Garden Weed Preventer Granules
Poast
Vegetable Turf & Ornamental
  Weeder
Weed Granules

PEPPER (Ornamental)
Poast

PERIWINKLE (Madagascar)
Amaze
Grass Out
Poast
Pre Emergent Weed & Grass
  Preventer
Weed & Grass Preventer

PERIWINKLE (Vinca)
Betasan 3.6
Eptam Weed Control
Eptam Weed & Grass Preventer
Grass-B-Gon
Poast
Surflan
Weed Stopper

PETUNIA
Azame
Eptam Weed Control
Eptam Weed & Grass Preventer
Garden Weed Preventer Granules
Grass-B-Gon
Grass Out
Poast
Pre Emergent Weed & Grass
  Preventer
Surflan
Vegetable Turf & Ornamental
  Weeder
Weed Granules
Weed & Grass Preventer
Weed Stopper

PHILODENDRON
Grass Out

PHOTINIA
Amaze
Casoron
Grass-B-Gon
Grass Out
Poast
Surflan
Weed & Grass Preventer
Weed Stopper

PIERIS (Andromeda)
Eptam Weed Control
Eptam Weed & Grass Preventer
Garden Weed Preventer Granules
Pre Emergent Weed & Grass
  Preventer
Surflan
Vegetable Turf & Ornamental
  Weeder
Weed Granules
Weed Stopper
Weed & Grass Preventer

PILEA (Creeping Charlie)
Grass Out

PINE
Amaze
Betasan 3.6
Eptam Weed Control
Eptam Weed & Grass Preventer
Garden Weed Preventer Granules
Grass-B-Gon
Poast
Pre Emergent Weed & Grass
  Preventer
Surflan
Vegetable Turf & Ornamental
  Weeder
Weed Granules
Weed & Grass Preventer
Weed Stopper

PINK CLOVER
Grass Out

PISTACHIO
Grass-B-Gon
Surflan
Weed Stopper

PITTOSPORUM
Amaze
Casoron
Garden Weed Preventer Granules
Grass-B-Gon
Grass Out
Poast
Surflan
Vegetable Turf & Ornamental
  Weeder

Weed Granules
Weed Stopper

PLAINTAIN LILY (Hosta)
Surflan
Weed Stopper

PLANT CONTAINERS
Lawn Moss Killer

PLUMS
Casoron
Grass-B-Gon
Poast
Surflan
Weed Stopper

PLUM (Flowering)
Grass-B-Gon

PLUMBAGO
Amaze
Grass Out
Poast
Weed Stopper

PODOCARPUS (Yew Pine)
Amaze
Eptam Weed Control
Eptam Weed & Grass Preventer
Garden Weed Preventer Granules
Grass Out
Poast
Pre Emergent Weed & Grass
  Preventer
Vegetable Turf & Ornamental
  Weeder
Weed Granules
Weed Stopper

POKER PLANT
Garden Weed Preventer Granules
Vegetable Turf & Ornamental
  Weeder
Weed Granules

POMEGRANATE
Poast
Surflan
Weed Stopper

PONDS/LAKES
Algae Attack
Copper Sulfate

POPLAR
Casoron
Garden Weed Preventer Granules
Poast
Vegetable Turf & Ornamental
  Weeder

HERBICIDES

USAGE OF PRODUCTS

POTATOES
Eptam Weed Control
Garden Turf & Ornamental Herbicide
Garden Weed Preventer Granules
Poast
Vegetable Turf & Ornamental Weeder

POTENILLA (Cinquefoil)
Amaze
Garden Weed Preventer Granules
Grass-B-Gon
Grass Out
Vegetable Turf & Ornamental Weeder
Weed Granules
Weed Stopper

POTHOS
Grass Out

PRIMROSE (Evening)
Betasan 3.6
Garden Weed Preventer Granules
Grass Out
Weed Granules

PRIVET (Ligustrum)
Amaze
Casoron
Garden Weed Preventer Granules
Grass-B-Gon
Grass Out
Poast
Surflan
Vegetable Turf & Ornamental Weeder
Weed Granules
Weed & Grass Preventer
Weed Stopper

PROTEA
Grass Out
Surflan
Weed Stopper

PRUNES
Casoron
Grass-B-Gon
Poast
Surflan
Weed Stopper

PUMPKINS
Poast

PURPLE HEART
Grass Out

PURPLELEAF (Acacia)
Poast

PURPLELEAF SAND CHERRY
Poast

PYRACANTHA (Firethorn)
Amaze
Betasan 3.6
Casoron
Grass Out
Poast
Surflan
Weed & Grass Preventer
Weed Stopper

QUINCE
Casoron
Poast

QUINCE (Flowering)
Grass-B-Gon

RADISHES
Garden Weed Preventer Granules
Vegetable Turf & Ornamental Weeder

RANUNCULUS
Betasan 3.6
Surflan
Weed Stopper

RASPBERRY ICE (Bougainvillea)
Grass Out
Poast

RED CEDAR (Thuja)
Surflan
Weed Stopper

RED FOUNTAIN GRASS
Grass Out
Poast

REDBUD
Amaze
Garden Weed Preventer Granules
Grass-B-Gon
Grass Out
Surflan
Vegetable Turf & Ornamental Weeder
Weed Granules
Weed Stopper

RED JUSTICIA
Grass Out

REDWOOD (Sequoia)
Amaze
Surflan
Weed Stopper

RHAPHIOLEPSIS (India Hawthorne/Pink Lady)
Amaze
Grass Out
Poast
Surflan
Weed Stopper

RHODENDRON (Azalea)
Casoron
Eptam Weed Control
Eptam Weed & Grass Preventer
Garden Weed Preventer Granules
Grass-B-Gon
Grass Out
Poast
Pre Emergent Weed & Grass Preventer
Surflan
Vegetable Turf & Ornamental Weeder
Weed Granules
Weed & Grass Preventer
Weed Stopper

ROCKROSE
Casoron
Grass Out

ROOFS
Moss Kil
Moss Kil Granules
Moss Kil Roof Strip
RTU Moss Kill

ROSES
Amaze
Casoron
Eptam Weed Control
Eptam Weed & Grass Preventer
Garden Weed Preventer Granules
Grass-B-Gon
Grass Out
Pre Emergent Weed & Grass Preventer
Rose Guard
Surflan
Vegetable Turf & Ornamental Weeder
Weed Granules
Weed Stopper

ROSE OF SHARON (Althea)
Amaze
Grass Out

85

HERBICIDES

Surflan
Weed Stopper

ROSEMARY
Amaze
Grass Out
Weed Stopper

RUBBER TREE (Ficus)
Grass Out

RUSSIAN OLIVE (Elaeagnus)
Amaze
Casoron
Garden Weed Preventer Granules
Grass Out
Poast
Surflan
Vegetable Turf & Ornamental Weeder
Weed Granules
Weed Stopper

SAGE (Scarlet)
Amaze
Garden Weed Preventer Granules
Surflan
Weed Granules

SALTBRUSH
Grass Out

SALVIA (Sage)
Poast
Weed Granules
Weed Stopper

SANDANKWA
Betasan 3.6

SANDWORT
Grass Out

SANSEVIERIA
Grass Out

SCHEFFLERA
Grass Out

SEA PINKS
Poast

SEDUM (Stonecrop)
Betasan 3.6
Eptam Weed Control
Eptam Weed & Grass Preventer
Garden Weed Preventer Granules
Grass Out
Poast
Pre Emergent Weed & Grass Preventer

Surflan
Vegetable Turf & Ornamental Weeder
Weed Stopper

SERVICEBERRY
Poast

SEWERS
Copper Sulfate

SHORT DAISY
Grass Out

SHRIMP PLANT (Justicia)
Amaze
Grass Out
Poast
Surflan
Weed Stopper

SHRUBS (general)
Amaze
Finale Concentrate
Finale RTU
Finale Super Concentrate
Grass & Weed Killer Kleen Up
Hose N Go Weed & Grass Preventer

SILVER KING
Poast

SILVER QUEEN
Grass Out

SKY FLOWER (Duranta)
Poast

SLIPPER FLOWER
Grass Out

SNAPDRADON
Amaze
Garden Weed Preventer Granules
Poast
Surflan
Vegetable Turf & Ornamental Weeder
Weed Granules
Weed Stopper

SNOWDRIFT
Amaze

SNOW IN SUMMER
Grass Out

SOTOL (Dasyiron)
Amaze
Surflan
Weed Stopper

SOUTHERN PEAS
Garden Turf & Ornamental Herbicide
Garden Weed Preventer Granules

SOYBEANS
Garden Weed Preventer Granules

SPIDER FLOWER
Grass Out

SPIDERWORT
Garden Weed Preventer Granules
Vegetable Turf & Ornamental Weeder

SPINACH
Poast

SPINDLE TREE
Poast

SPIREA (Bridal Wreath)
Amaze
Casoron
Garden Weed Preventer Granules
Grass-B-Gon
Grass Out
Poast
Vegetable Turf & Ornamental Weeder
Weed Granules
Weed Stopper

SPRUCE
Amaze
Eptam Weed Control
Eptam Weed & Grass Preventer
Garden Weed Preventer Granules
Grass-B-Gon
Grass Out
Poast
Pre Emergent Weed & Grass Preventer
Surflan
Vegetable Turf & Ornamental Weeder
Weed Granules
Weed & Grass Preventer
Weed Stopper

SQUASH
Garden Turf & Ornamental Herbicide
Garden Weed Preventer Granules
Poast
Vegetable Turf & Ornamental Weeder

SQUAW CARPET
Casoron

HERBICIDES

STAR JASMINE
Betasan 3.6
Grass-B-Gon
Grass Out
Poast
Surflan
Weed & Grass Preventer
Weed Stopper

STAR PLANT
Grass Out
Poast

STATICE
Poast

STOCK
Betasan 3.6

STOKES ASTER
Weed Stopper

STORE CROP
Amaze

STRAWBERRIES
Garden Turf & Ornamental Herbicide
Garden Weed Preventer Granules
Poast
Vegetable Turf & Ornamental Weeder
Weed Granules

STRAWBERRY (Ornamental)
Betasan 3.6
Eptam Weed Control
Eptam Weed & Grass Preventer
Grass Out
Pre Emergent Weed & Grass Preventer

STRAWBERRY TREE (Arbutus)
Grass Out
Poast

STRAWFLOWER
Garden Weed Preventer Granules
Weed Granules

STUMP TREATMENT
Brush-B-Gon
Brushkil
Brush Buster
Brush Killer
Roundup

SUMAC
Amaze
Grass Out
Poast

SUMAC (Rhea)
Surflan
Weed Stopper

SUNDROPS
Garden Weed Preventer Granules
Vegetable Turf & Ornamental Weeder
Weed Granules

SUNFLOWER
Garden Weed Preventer
Vegetable Turf & Ornamental Weeder
Weed Granules

SWEETGUM
Grass Out
Weed & Grass Preventer

SWEET GRASS
Poast

SWEET PEAS
Betasan 3.6
Garden Weed Preventer Granules
Vegetable Turf & Ornamental Weeder
Weed Granules

SWEET POTATO (Yams)
Garden Turf & Ornamental Herbicide
Garden Weed Preventer Granules
Vegetable Turf & Ornamental Weeder

SWEET WILLIAM (Dianthus)
Amaze
Grass Out
Poast
Surflan
Weed Stopper

SYCAMORE
Amaze
Garden Weed Preventer Granules
Poast
Vegetable Turf & Ornamental Weeder
Weed Granules
Weed Stopper

TALLHEDGE BUCKTHORN
Grass Out
Poast

TEA TREE
Poast

TEMPLE TREE
Grass Out

TEXAS SAGE
Grass Out

THUMBERGIA
Grass Out

TIPU TREE
Poast

TOBACCO (Flowering)
Poast

TOBIRA (Pittosporum)
Betasan 3.6
Surflan
Weed Stopper

TOMATOES
Garden Turf & Ornamental Herbicide
Garden Weed Preventer Granules
Poast
Vegetable Turf & Ornamental Weeder
Weed Granules

TOYON
Amaze
Poast
Weed Stopper

TREES (General)
Finale Concentrate
Finale RTU
Finale Super Concentrate
Grass & Weed Killer Kleen Up RTC
Home Deck & Patio Moss & Algae Killer
Hose N Go Weed & Grass Preventer

TREE PEONY
Garden Weed Preventer Granules
Vegetable Turf & Ornamental Weeder
Weed Granules

TRUMPET VINE
Amaze
Poast
Surflan
Weed Stopper

TULIP TREE
Garden Weed Preventer Granules
Grass-B-Gon
Weed Granules

USAGE OF PRODUCTS

HERBICIDES

TULIP
- Betasan 3.6
- Surflan
- Weed & Grass Preventer
- Weed Stopper

TURF (general)
- 2,4-D Amino Weed Killer
- Balfin Granules
- Broadleaf Weed & Dandelion Control
- Crabgrass Control & Lawn Food
- Crabgrass & Nutgrass Killer
- Crabgrass Killer
- Crabgrass Plus Broadleaf Weed Killer
- Crabgrass Preventer & Weed Killer
- Dandelion & Broadleaf Weed Control Conc.
- Dandelion Killer
- Ezy Spray Weed & Feed
- First Down
- Garden Turf & Ornamental Herbicide
- Green Sweep Weed & Feed
- Home Deck & Patio Moss & Algae Killer
- Hose-n-Go Moss Out
- Hose & Go Weed & Feed
- Lawn Food & Weed Control
- Lawn Food Plus Moss Control
- Lawn Moss Killer
- Lawn Spot Weeder
- Lawn Weed Killer
- Liquid Weed & Feed 15-2-3
- Moss Out
- Moss Out Granules
- Moss Out Plus Fertilizers
- MSMA Crabgrass Killer
- Poison Ivy & Brush Killer
- Protrait
- Rapid Green Weed & Feed
- RTU Spurge Oxalis & Dandelion Killer
- RTU Lawn Weed Killer
- Spot Weed Killer
- Spot Weeder
- Spurge & Oxalis Killer
- Super Chickweed Killer
- Super Rich Lawn Food with Moss Control
- Super Rich Weed & Feed
- Supreme Lawn Fertilizer & rabgrass Preventer
- Triamine Weed & Feed
- Trimec Lawn Weed Killer
- Trimec Plus
- Trimec Weed & Feed 24-4-8
- Turf & Ornamental Herb.
- Ultra Green Moss Control Lawn Food
- Ultra Green Weed & Feed
- Vegetable Turf & Ornamental Weeder
- Weed & Feed
- Weed & Feed 12-2-4
- Weed & Feed Granule 17-4-4
- Weed-B-Gon
- Weed-B-Gon for Southern Lawns
- Weed-B-Gon Jet Weeder
- Weed-B-Gon Lawn Weed Killer 2
- Weed-B-Gon Ready Spray
- Weed-No-More Spot Weeder
- Weed-Out Lawn Weed Killer
- Weed Wacker
- Weed Wacker Jet Spray
- Weed Warrior
- Weed Warrior Spot Weed Killer
- Wipe-Out
- Wipe-Out RTU
- Yard Basics Lawn Weed & Feed
- Yard Basics Weed Killer for Lawns

TURF (Cool Season)
- Garden Weed Preventer Granules
- Weed Hoe
- Weed Whacker

TURF (Newly Seeded)
- Crabgrass Preventer & Weed Killer
- Garden Weed Preventer Granules
- Turf & Ornamental Herb.
- Vegetable Turf & Ornamental Weeder

TURF (Renovation)
- Kleen Away
- Roundup
- Weed Ender

TURF (Southern)
- Weed & Grass Preventer

TURF (Warm Season)
- Amaze
- Atrazine 4L
- Garden Weed Preventer Granules
- Weed Hoe
- Weed Stopper
- Weed Whacker

TURNIPS
- Garden Turf & Ornamental Herbicide
- Garden Weed Preventer Granules
- Vegetable Turf & Ornamental Weeder
- Weed Granules

UMBRELLA PLANT
- Grass Out

VEGETABLES (before planting)
- Klearaway
- Roundup

VERBENA
- Poast

VERONICA
- Poast
- Weed Granules

VIBURNUM (Snowball)
- Amaze
- Eptam Weed Control
- Eptam Weed & Grass Preventer
- Garden Weed Preventer Granules
- Grass-B-Gon
- Grass Out
- Poast
- Pre Emergent Weed & Grass Preventer
- Surflan
- Vegetable Turf & Ornamental Weeder
- Weed Granules
- Weed & Grass Preventer
- Weed Stopper

VIOLET (African)
- Garden Weed Preventer Granules
- Vegetable Turf & Ornamental Weeder

VIOLETS
- Weed Granules

WALL FLOWER
- Betasan 3.6

WALNUT
- Casoron
- Garden Weed Preventer Granules
- Grass-B-Gon
- Poast
- Surflan
- Vegetable Turf & Ornamental Weeder
- Weed Granules
- Weed Stopper

WANDERING JEW
- Grass Out
- Poast

WATERMELON
- Garden Turf & Ornamental Herbicide

HERBICIDES

Garden Weed Preventer Granules
Poast
Vegetable Turf & Ornamental
 Weeder

WAX MYRTLE
Grass Out

WEIGELA
Amaze
Casoron
Garden Weed Preventer Granules
Grass Out
Surflan
Vegetable Turf & Ornamental
 Weeder
Weed Granules
Weed Stopper

WHEELERS DWARF (Pittsporium)
Poast

WILLOW
Amaze
Casoron
Garden Weed Preventer Granules
Grass-B-Gon
Grass Out
Poast
Vegetable Turf & Ornamental
 Weeder
Weed Granules
Weed Stopper

WILLOW (Desert)
Poast

WILLOW (Peppermint)
Poast

WINTERCREEPER
Amaze
Surflan
Weed Stopper

WOADWAXEN (Genista)
Amaze
Weed Stopper

WOODBINE
Amaze
Weed Stopper

WORMWOOD
Garden Weed Preventer Granules
Vegetable Turf & Ornamental
 Weeder
Weed Granules

XYLOSMA
Amaze

Surflan
Weed & Grass Preventer
Weed Stopper

YARROW (Achilea)
Amaze
Garden Weed Preventer Granules
Grass Out
Surflan
Vegetable Turf & Ornamental
 Weeder
Weed Granules
Weed Stopper

YAUPON
Grass Out
Surflan
Weed Stopper

YELLOW BELLS (Tecoma)
Grass Out
Poast

YELLOW TAB (Tabebuia)
Amaze
Weed Stopper

YELLOW TRUMPET
Grass Out

YEW (Taxus)
Amaze
Casoron
Eptam Weed Control
Eptam Weed & Grass Preventer
Garden Weed Preventer Granules
Grass-B-Gon
Grass Out
Poast
Pre Emergent Weed & Grass
 Preventer
Surflan
Vegetable Turf & Ornamental
 Weeder
Weed Granules
Weed Stopper

YUCCA
Amaze
Grass Out
Surflan
Weed Stopper

ZINNIA
Amaze
Betasan 3.6
Eptam Weed Control
Eptam Weed & Grass Preventer
Garden Weed Preventer Granules
Grass-B-Gon
Poast
Pre Emergent Weed & Grass

Preventer
Surflan
Vegetable Turf & Ornamental
 Weeder
Weed Granules
Weed & Grass Preventer
Weed Stopper

FUNGICIDES

AFRICAN VIOLET
Bayleton
Dithane M-45
Fungicide
Lawn, Ornamental & Vegetable
   Fungicide
Mancozeb
Mancozeb Plant Fungicide
Procide
Systemic Fungicide

AGERATUM
Bayleton
Orthenex Insect & Disease Control I
Procide
Systemic Fungicide

AGLAENEMA
Lawn, Ornamental & Vegetable
   Fungicide

ALMOND
Captan
Copper Spray or Dust
Daconil
Kop-R-Spray
Lawn, Ornamental & Vegetable
   Fungicide
Liqui-Cop
Sulfur Dust
Systemic Fungicide
Wettable Dusting Sulfur

ALMOND (Flowering)
Bravo
Daconil 2787
Multi-Purpose Fungicide
Orthenex Insect & Disease Control I
Systemic Fungicide 3336

AMELANCHIER
Bayleton
Procide
Systemic Fungicide

ANTHURIUM
Dithane M-45
Mancozeb

APPLES
Bayleton
Bordeaux Mix
Bordeaux Mixture
Bordeaux Powder
Captan
Captan 50% WP
Captan Wettable
Complete Fruit Tree Spray
Copper Fungicide
Copper Spray or Dust
Dormant Disease Control
Ferbam
Flower Fruit & Vegetable Garden
   Fungicide
Fruit Spray Concentrate
Fruit Tree Spray
Foli-Cal
Fruit Tree Spray
Garden Sulpher
Home Orchard Spray
Kop-R-Spray
Lime Sulfur Solution
Lime Sulfur Spray
Liquid Fruit Tree Spray
Liquid Sulfur
Microcop
Polysul
Sulf-R-Spray
Sulfur Dust
Sulfur Plant Fungicide
Systemic Fungicide
Systemic Fungicide 3336
Wettable Dusting Sulfur

APRICOTS
Bordeaux Mixture
Bravo
Captan
Copper Fungicide
Copper Spray or Dust
Daconil
Daconil 2787
Fruit Tree Spray
Home Orchard Spray
Kop-R-Spray
Lawn, Ornamental & Vegetable
   Fungicide
Liqui-Cop
Microcop
Multi-Purpose Fungicide
Systemic Fungicide
Systemic Fungicide 3336

ARALIA
Lawn, Ornamental & Vegetable
   Fungicide

ARBORVITAE
Bordeaux Mix
Bordeaux Mixture
Copper Fungicide
Copper Spray or Dust
Dithane M-45
Garden Sulphur
Kop-R-Spray
Mancozeb
Mancozeb Plant Fungicide
Microcop

ARTEMESIA
Lawn, Ornamental & Vegetable
   Fungicide

ARTISIA
Orthenex Insect & Disease Control I

ASH
Bayleton
Bravo
Daconil
Daconil 2787
Lawn, Ornamental & Vegetable
   Fungicide
Lime Sulfur Solution
Mancozeb
Mancozeb Plant Fungicide
Multi-Purpose Fungicide
Procide

ASPARAGUS
Copper Dragon
Mancozeb
Mancozeb Fungicide
Mancozeb Plant Fungicide

ASPEN
Bayleton
Procide
Systemic Fungicide

ASTER
Bayleton
Copper Spray or Dust
Dithane M-45
Funginex
Garden Sulphur
Kop-R-Spray
Mancozeb
Mancozeb Plant Fungicide
Orthenex Insect & Disease Control
Orthenex Insect & Disease Control I
Procide
Systemic Fungicide
Wettable Dusting Sulfur

AVOCADOS
Copper Fungicide
Kop-R-Spray
Liquid Copper
Wettable Dusting Sulfur

AZALEA
Bayleton
Bordeaux Mix
Bordeaux Mixture
Bravo
Captan
Captan 50% WP
Captan Wettable
Copper Dragon
Copper Fungicide
Copper Spray or Dust
Daconil
Daconil 2787

FUNGICIDES

USAGE OF PRODUCTS

Disease Control
Dithane M-45
Fungicide
Funginex
Garden Sulphur
Kop-R-Spray
Lawn, Ornamental & Vegetable Fungicide
Liquid Copper
Mancozeb
Microcop
Multi-Purpose Fungicide
Orthenex Insect & Disease Control
Procide
Rose & Ornamental Fungicide
Systemic Fungicide
Systemic Fungicide 3336
Thiomyl
Wettable Dusting Sulfur

BALD CYPRESS
Orthenex Insect & Disease Control I

BALSAM PEAR
Bayleton
Systemic Fungicide

BANANAS
Mancozeb Fungicide

BARBERRY
Bayleton
Bordeaux Mix
Bordeaux Mixture
Kop-R-Spray
Procide
Systemic Fungicide

BEANS
Bravo
Captan 50% WP
Copper Dragon
Copper Fungicide
Daconil
Daconil 2787
Flower Fruit & Vegetable Garden Fungicide
Kop-R-Spray
Lawn, Ornamental & Vegetable Fungicide
Liquid Copper
Liquid Copper Fungicide
Multi-Purpose Fungicide
Sulfur Dust
Sulfur Plant Fungicide
Tomato Potato Dust
Wettable Dusting Sulfur

BEDDING PLANT (general)
Flower Fruit & Vegetable Garden Fungicide
Fungicide
Rose & Ornamental Fungicide
Systemic Fungicide 3336

BEETS
Captan 50% WP
Copper Fungicide
Liquid Copper

BEGONIA
Bayleton
Captan
Captan 50% WP
Copper Fungicide
Copper Spray or Dust
Dithane M-45
Dormant Disease Control
Fungicide
Funginex
Kop-R-Spray
Lawn, Ornamental & Vegetable Fungicide
Liquid Copper
Mancozeb
Mancozeb Fungicide
Mancozeb Plant Fungicide
Procide
Systemic Fungicide

BIRCH
Bayleton
Lime Sulfur Solution
Orthenex Insect & Disease Control I
Procide
Systemic Fungicide

BLUEBERRIES
Wettable Dusting Sulfur

BOXWOOD
Bordeaux Mix
Bordeaux Mixture
Kop-R-Spray
Lime Sulfur Solution

BROCCOLI
Bravo
Copper Dragon
Copper Fungicide
Copper Spray or Dust
Daconil
Daconil 2787
Fungicide
Kop-R-Spray
Lawn & Garden Fungicide
Lawn, Ornamental & Vegetable Fungicide
Liquid Copper
Maneb Tomato & Vegetable Fungicide
Multi-Purpose Fungicide
Tomato Potato Dust
Wettable Dusting Sulfur

BRUSSELS SPROUTS
Bravo
Copper Fungicide
Copper Spray or Dust
Daconil
Daconil 2787
Fungicide
Kop-R-Spray
Lawn & Garden Fungicide
Lawn, Ornamental & Vegetable Fungicide
Liquid Copper
Maneb Tomato & Vegetable Fungicide
Multi-Purpose Fungicide
Tomato Potato Dust
Wettable Dusting Sulfur

BUCKEYE (Horse chestnut)
Bayleton
Bravo
Daconil 2787
Dithane M-45
Lawn, Ornamental & Vegetable Fungicide
Mancozeb
Multi-Purpose Fungicide
Procide
Systemic Fungicide

BUCKTHORN
Bayleton
Procide
Systemic Fungicide

BUFFALOBERRY
Mancozeb

CABBAGE
Bravo
Captan 50% WP
Copper Dragon
Copper Fungicide
Copper Spray or Dust
Daconil
Daconil 2787
Fungicide
Kop-R-Spray
Lawn & Garden Fungicide
Lawn, Ornamental & Vegetable Fungicide
Liquid Copper
Maneb Tomato & Vegetable Fungicide
Multi-Purpose Fungicide
Tomato Potato Dust
Wettable Dusting Sulfur

CALENDULA
Bayleton
Funginex

FUNGICIDES

Garden Sulphur
Orthenex Insect & Disease Control
Orthenex Insect & Disease Control I
Procide
Systemic Fungicide
Wettable Dusting Sulfur

CAMELLIA

Bayleton
Bordeaux Mixture
Captan
Captan Wettable
Copper Fungicide
Dithane M-45
Fungicide
Kop-R-Spray
Lawn, Ornamental & Vegetable
  Fungicide
Liquid Copper
Mancozeb
Microcop
Orthenex Insect & Disease Control I
Procide
Systemic Fungicide

CANNA

Bayleton
Procide
Systemic Fungicide

CANEBERRIES

Copper Fungicide
Dormant Disease Control
Kop-R-Spray
Lime Sulfur Solution
Lime Sulfur Spray
Liquid Copper
Polysul
Sulfur Dust
Sulfur Plant Fungicide
Sulf-R-Spray
Wettable Dusting Sulfur

CARNATION (Dianthus)

Bayleton
Bravo
Captan
Captan 50% WP
Daconil 2787
Dithane M-45
Ferbam
Fungicide
Funginex
Garden Sulphur
Kop-R-Spray
Lawn, Ornamental & Vegetable
  Fungicide
Mancozeb
Mancozeb Fungicide
Mancozeb Plant Fungicide
Maneb Tomato & Vegetable
  Fungicide

Multi-Purpose Fungicide
Orthenex Insect & Disease Control
Orthenex Insect & Disease Control I
Procide
Systemic Fungicide
Wettable Dusting Sulfur

CARROTS

Bravo
Copper Dragon
Copper Fungicide
Copper Spray or Dust
Daconil 2787
Kop-R-Spray
Lawn, Ornamental & Vegetable
  Fungicide
Liquid Copper
Liquid Copper Fungicide
Multi-Purpose Fungicide
Sulfur Dust
Wettable Dusting Sulfur

CATALPA

Systemic Fungicide 3336

CAULIFLOWER

Bravo
Copper Fungicide
Copper Spray or Dust
Daconil
Daconil 2787
Fungicide
Kop-R-Spray
Lawn & Garden Fungicide
Lawn, Ornamental & Vegetable
  Fungicide
Liquid Copper
Multi-Purpose Fungicide
Tomato Potato Dust
Wettable Dusting Sulfur

CEDAR

Bayleton
Bordeaux Mixture
Copper Fungicide
Dithane M-45
Garden Sulphur
Kop-R-Spray
Liquid Copper
Mancozeb
Mancozeb Plant Fungicide
Orthenex Insect & Disease Control I

CELERY

Bordeaux Mix
Bordeaux Mixture
Bordeaux Powder
Bravo
Copper Fungicide
Copper Spray or Dust
Daconil 2787
Foli-Cal

Kop-R-Spray
Lawn, Ornamental & Vegetable
  Fungicide
Liquid Copper
Microcop
Multi-Purpose Fungicide

CHERRY

Bordeaux Mixture
Bordeaux Powder
Bravo
Captan
Captan Wettable
Complete Fruit Tree Spray
Copper Fungicide
Copper Spray or Dust
Daconil
Daconil 2787
Fruit Spray Concentrate
Fruit Tree Spray
Garden Sulpher
Home Orchard Spary
Kop-R-Spray
Lawn, Ornamental & Vegetable
  Fungicide
Lime Sulfur Spray
Liqui-Cop
Liquid Copper
Liquid Fruit Tree Spray
Liquid Sulfur
Multi-Purpose Fungicide
Polysul
Sulf-R-Spray
Sulfur Dust
Sulfur Plant Fungicide
Systemic Fungicide
Systemic Fungicide 3336
Wettable Dusting Sulfur

CHERRY LAUREL

Bravo
Daconil
Daconil 2787
Lawn, Ornamental & Vegetable
  Fungicide
Multi-Purpose Fungicide
Orthenex Insect & Disease Control I

CHESTNUT

Bayleton
Bravo
Procide
Systemic Fungicide

CHINESE CABBAGE

Lawn, Ornamental & Vegetable
  Fungicide
Maneb Tomato & Vegetable
  Fungicide

CHINESE CURCURBITS

Bayleton

FUNGICIDES

CHRISTMAS TREES (General)
Bravo

CHRYSANTHEMUM
Bayleton
Bordeaux Mix
Bordeaux Mixture
Bravo
Captan
Captan 50% WP
Captan Wettable
Copper Fungicide
Copper Spray or Dust
Daconil
Daconil 2787
Dithane M-45
Ferbam
Fungicide
Garden Sulphur
Kop-R-Spray
Lawn & Garden Fungicide
Lawn, Ornamental & Vegetable Fungicide
Liquid Copper
Mancozeb
Mancozeb Plant Fungicide
Multi-Purpose Fungicide
Orthenex Insect & Disease Control I
Procide
Sulfur Dust
Systemic Fungicide
Wettable Dusting Sulfur

CINERARIA
Bayleton
Procide
Systemic Fungicide

CITRON MELON
Bayleton

CITRUS
Bordeaux Mixture
Copper Fungicide
Dormant Disease Control
Flower Fruit & Vegetable Garden Fungicide
Kop-R-Spray
Liqui-Cop
Liquid Copper
Liquid Copper Spray
Liquid Sulfur
Sulfur Dust
Sulfur Plant Fungicide
Wettable Dusting Sulfur

CLEMATIS
Garden Sulphur
Sulfur Dust

COLEUS
Orthenex Insect & Disease Control I

COLLARDS
Copper Dragon
Wettable Dusting Sulfur

COLUMBINE
Dormant Disease Control

CONIFERS (General)
Bravo
Daconil
Dithane M-45
Lawn, Ornamental & Vegetable Fungicide
Mancozeb
Mancozeb Plant Fungicide
Multi-Purpose Fungicide
Rose & Ornamental Fungicide
Systemic Fungicide
Systemic Fungicide 3336

CORDYLINE
Dithane M-45
Mancozeb

CORN (Sweet)
Captan 50% WP
Daconil 2787
Dithane M-45
Lawn, Ornamental & Vegetable Fungicide
Mancozeb
Mancozeb Fungicide
Mancozeb Plant Fungicide
Maneb Tomato & Vegetable Fungicide
Multi-Purpose Fungicide
Tomato Potato Dust

COTONEASTER
Copper Fungicide
Liquid Copper

COTTONWOOD
Bayleton
Procide

CRABAPPLE
Bayleton
Bravo
Daconil
Daconil 2787
Dithane M-45
Lawn, Ornamental & Vegetable Fungicide
Mancozeb
Mancozeb Plant Fungicide
Multi-Purpose Fungicide
Orthenex Insect & Disease Control I
Procide
Rose & Ornamental Fungicide
Systemic Fungicide
Systemic Fungicide 3336

CRANBERRIES
Mancozeb Fungicide

CRAPE MYRTLE
Bayleton
Dormant Disease Control
Funginex
Garden Sulphur
Orthenex Insect & Disease Control
Orthenex Insect & Disease Control I
Procide
Systemic Fungicide
Wettable Dusting Sulfur

CRASSULA
Bayleton
Procide
Systemic Fungicide

CROSUS
Lawn, Ornamental & Vegetable Fungicide

CROTON
Orthenex Insect & Disease Control I

CUCUMBERS
Bayleton
Bordeaux Powder
Bravo
Copper Dragon
Copper Fungicide
Copper Spray or Dust
Daconil
Daconil 2787
Dithane M-45
Flower Fruit & Veg Garden Fungicide
Kop-R-Spray
Lawn, Ornamental & Vegetable Fungicide
Liquid Copper
Liquid Copper Fungicide
Mancozeb
Maneb Tomato & Vegetable Fungicide
Multi-Purpose Fungicide
Tomato Potato Dust
Wettable Dusting Sulfur

CURRANTS
Bordeaux Powder
Lime Sulfur Solution
Lime Sulfur Spray
Sulfur Plant Fungicide
Wettable Dusting Sulfur

93

FUNGICIDES

CYPRESS
Bordeaux Mix
Bordeaux Mixture
Copper Fungicide
Kop-R-Spray
Mancozeb
Microcop

DAFODIL
Lawn, Ornamental & Vegetable Fungicide
Rose & Ornamental Fungicide
Systemic Fungicide 3336

DAHLIA
Bayleton
Bordeaux Mix
Bordeaux Mixture
Dithane M-45
Funginex
Garden Sulphur
Kop-R-Spray
Mancozeb
Mancozeb Fungicide
Maneb Tomato & Vegetable Fungicide
Orthenex Insect & Disease Control
Orthenex Insect & Disease Control I
Procide
Systemic Fungicide
Wettable Dusting Sulfur

DAISY
Bayleton
Daconil
Daconil 2787
Lawn, Ornamental & Vegetable Fungicide
Multi-Purpose Fungicide
Systemic Fungicide

DELPHINIUM
Bayleton
Bordeaux Mixture
Dormant Disease Control
Funginex
Garden Sulphur
Kop-R-Spray
Procide
Systemic Fungicide
Wettable Dusting Sulfur

DENDROBIUM
Bayleton
Procide
Systemic Fungicide

DESMELLA
Systemic Fungicide

DEWBERRIES
Sulfur Plant Fungicide
Wettable Dusting Sulfur

DICHONDRA
Bravo
Daconil 2787
Fungicide
Lawn & Garden Fungicide
Multi-Purpose Fungicide

DIEFFENBACHIA
Dithane M-45
Lawn, Ornamental & Vegetable Fungicide
Mancozeb
Mancozeb Plant Fungicide

DOGWOOD
Bayleton
Bordeaux Mix
Bordeaux Mixture
Bravo
Copper Fungicide
Daconil
Disease Control
Dithane M-45
Garden Sulphur
Kop-R-Spray
Lawn, Ornamental & Vegetable Fungicide
Mancozeb
Mancozeb Fungicide
Maneb Tomato & Vegetable Fungicide
Microcop
Multi-Purpose Fungicide
Procide
Systemic Fungicide
Wettable Dusting Sulfur

DOUGLAS FIR
Daconil 2787
Mancozeb
Multi-Purpose Fungicide

DRACAERA
Bravo
Daconil 2787
Dithane M-45
Lawn, Ornamental & Vegetable Fungicide
Mancozeb
Mancozeb Plant Fungicide
Multi-Purpose Fungicide

EASTER LILY
Rose & Ornamental Fungicide
Systemic Fungicide 3336

ELM
Bayleton
Bordeaux Mix
Bordeaux Mixture
Bordeaux Powder
Dithane M-45
Kop-R-Spray
Lime Sulfur Solution
Mancozeb
Mancozeb Plant Fungicide
Procide
Systemic Fungicide

EGGPLANT
Copper Dragon
Kop-R-Spray
Maneb Tomato & Vegetable Fungicide
Tomato Potato Dust

ENDIVE
Maneb Tomato & Vegetable Fungicide

EUCALYPTUS
Lawn, Ornamental & Vegetable Fungicide

EUONYMOUS
Bayleton
Bravo
Daconil
Daconil 2787
Dithane M-45
Dormant Disease Control
Funginex
Lawn, Ornamental & Vegetable Fungicide
Lime Sulfur Solution
Mancozeb
Mancozeb Fungicide
Mancozeb Plant Fungicide
Multi-Purpose Fungicide
Orthenex Insect & Disease Control
Orthenex Insect & Disease Control I
Procide
Systemic Fungicide
Wettable Dusting Sulfur

EVERGREENS (general)
Complete Fruit Tree Spray
Daconil
Fruit Tree Spray
Liquid Fruit Tree Spray
Rose & Flower Spray or Dust
Rose & Flower Insect & Disease Concentrate

FATSIA (Aralia)
Dithane M-45

FUNGICIDES

Lawn, Ornamental & Vegetable Fungicide
Mancozeb

FENNEL
Mancozeb

FERN (Boston)
Bayleton
Lawn, Ornamental & Vegetable Fungicide
Procide
Systemic Fungicide

FERN (general)
Dithane M-45
Lawn, Ornamental & Vegetable Fungicide
Mancozeb
Mancozeb Plant Fungicide

FERN (Leatherleaf)
Bravo
Daconil
Daconil 2787
Lawn, Ornamental & Vegetable Fungicide
Multi-Purpose Fungicide

FICUS
Dithane M-45
Lawn, Ornamental & Vegetable Fungicide
Mancozeb
Mancozeb Plant Fungicide
Orthenex Insect & Disease Control I

FILBERTS
Copper Fungicide
Kop-R-Spray
Microcop

FIR
Bayleton
Bravo
Dithane M-45
Lawn, Ornamental & Vegetable Fungicide
Mancozeb Plant Fungicide
Procide

FLOWERS (general)
Complete Fruit Tree Spray
Flower Fruit & Vegetable Garden Fungicide
Fruit Tree Spray
Fung-Away
Liquid Fruit Tree Spray
Orthenex Insect & Disease Control
Polysul
Rose & Flower Spray or Dust
Rose & Ornamental Fungicide

Rose & Flower Insect & Disease Concentrate
Sulfur Dust
Sulf-R-Spray
Systemic Fungicide
Systemic Fungicide 3336
Thiomyl

FLOWER SEED BEDS
Captan

FOLIAGE PLANTS (general)
Copper Fungicide
Flower Fruit & Vegetable Garden Fungicide
Liquid Copper

FOUR O'CLOCK
Bayleton
Procide
Systemic Fungicide

FRUIT TREES (general)
Dormant Disease Control
Polysul
Sulf-R-Spray

FUCHSIA
Dithane M-45
Mancozeb
Mancozeb Plant Fungicide
Orthenex Insect & Disease Control I

GARDENIA
Bayleton
Copper Fungicide
Copper Spray or Dust
Kop-R-Spray
Liquid Copper
Orthenex Insect & Disease Control I
Procide
Systemic Fungicide

GARLIC
Bravo
Daconil 2787
Lawn, Ornamental & Vegetable Fungicide
Multi-Purpose Fungicide
Wettable Dusting Sulfur

GERANIUM
Bayleton
Bordeaux Mix
Bordeaux Mixture
Bravo
Daconil
Daconil 2787
Disease Control
Dithane M-45
Fungicide
Kop-R-Spray

Lawn, Ornamental & Vegetable Fungicide
Lawn & Garden Fungicide
Mancozeb
Mancozeb Plant Fungicide
Multi-Purpose Fungicide
Procide
Systemic Fungicide

GERBERA
Bayleton
Procide
Systemic Fungicide

GLADIOLUS
Bordeaux Mix
Bordeaux Mixture
Bravo
Bulb Dust
Captan
Captan 50% WP
Daconil
Daconil 2787
Dithane M-45
Kop-R-Spray
Lawn, Ornamental & Vegetable Fungicide
Maneb Tomato & Vegetable Fungicide
Mancozeb
Mancozeb Fungicide
Mancozeb Plant Fungicide
Multi-Purpose Fungicide
Rose and Ornamental Fungicide
Systemic Fungicide 3336

GOLD DUST PLANT
Dithane M-45

GOOSEBERRIES
Bordeaux Powder
Lime Sulfur Solution
Lime Sulfur Spray
Sulfur Dust
Wettable Dusting Sulfur

GOURDS
Bayleton

GRAPE LEAF IVY
Bayleton

GRAPES
Bayleton
Bordeaux Mix
Bordeaux Mixture
Bordeaux Powder
Captan
Captan 50% WP
Captan Wettable
Complete Fruit Tree Spray
Copper Dragon

USAGE OF PRODUCTS

FUNGICIDES

Copper Fungicide
Dormant Disease Control
Ferbam
Foli-Cal
Fruit Spray Concentrate
Fruit Tree Spray
Garden Sulphur
Home Orchard Spray
Kop-R-Spray
Lime Sulfur Spray
Liquid Copper
Liquid Copper Spray
Liquid Fruit Tree Spray
Liquid Sulfur
Mancozeb Fungicide
Maneb Tomato & Vegetable Fungicide
Sulfur Dust
Sulfur Plant Fungicide
Wettable Dusting Sulfur

GREENHOUSES (home)

Flower Fruit & Vegetable Garden Fungicide
Thiomyl

GROUND COVERS (general)

Thiomyl

HAWTHORN

Bayleton
Bravo
Copper Spray or Dust
Daconil
Daconil 2787
Dithane M-45
Kop-R-Spray
Lawn, Ornamental & Vegetable Fungicide
Mancozeb
Mancozeb Plant Fungicide
Multi-Purpose Fungicide
Procide
Systemic Fungicide

HEDGES (general)

Lime Sulfur Spray

HEMLOCK

Bayleton

HOLLY

Bayleton
Bravo
Daconil 2787
Dithane M-45
Lawn, Ornamental & Vegetable Fungicide
Mancozeb
Mancozeb Plant Fungicide
Multi-Purpose Fungicide
Procide
Systemic Fungicide
Wettable Dusting Sulfur

HOLLYHOCK

Bayleton
Bordeaux Mix
Bordeaux Mixture
Bravo
Daconil
Daconil 2787
Dithane M-45
Kop-R-Spray
Lawn, Ornamental & Vegetable Fungicide
Mancozeb
Mancozeb Fungicide
Multi-Purpose Fungicide
Procide
Sulfur Dust
Systemic Fungicide
Wettable Dusting Sulfur

HONEYSUCKLE

Dithane M-45
Mancozeb
Mancozeb Plant Fungicide

HUCKLEBERRIES

Wettable Dusting Sulfur

HYACINTH

Bulb Dust
Fungicide

HYDRANGEA

Bravo
Copper Fungicide
Daconil
Daconil 2787
Dithane M-45
Garden Sulphur
Lawn, Ornamental & Vegetable Fungicide
Liquid Copper
Mancozeb
Mancozeb Fungicide
Mancozeb Plant Fungicide
Multi-Purpose Fungicide
Systemic Fungicide 3336

IRIS

Bayleton
Bordeaux Mix
Bordeaux Mixture
Bravo
Bulb Dust
Copper Fungicide
Daconil
Daconil 2787
Dithane M-45
Fungicide
Kop-R-Spray
Lawn & Garden Fungicide
Lawn, Ornamental & Vegetable Fungicide
Liquid Copper
Mancozeb
Mancozeb Plant Fungicide
Multi-Purpose Fungicide
Rose & Ornamental Fungicide
Procide
Systemic Fungicide
Systemic Fungicide 3336

IVY

Bordeaux Mix
Bordeaux Mixture
Copper Fungicide
Kop-R-Spray
Liquid Copper
Orthene Insect & Disease Control I
Procide
Systemic Fungicide

IXORA

Liquid Copper

JERUSALEM THORN

Funginex
Orthenex Insect & Disease Control

JUNIPER

Bayleton
Bordeaux Mix
Bordeaux Mixture
Bravo
Copper Fungicide
Dithane M-45
Kop-R-Spray
Lime Sulfur Solution
Liquid Copper
Mancozeb
Mancozeb Plant Fungicide
Microcop
Orthene Insect & Disease Control I
Procide
Systemic Fungicide 3336

KALANCHOE

Bayleton
Funginex
Procide
Systemic Fungicide

KALE

Copper Dragon
Wettable Dusting Sulfur

KOHLRABI

Tomato Potato Dust

LAUREL

Bordeaux Mix
Bordeaux Mixture

FUNGICIDES

USAGE OF PRODUCTS

Copper Fungicide
Kop-R-Spray
Mancozeb
Microcop

LEEKS
Bravo
Daconil 2787
Lawn, Ornamental & Vegetable Fungicide
Multi-Purpose Fungicide

LETTUCE
Copper Dragon
Liquid Copper Fungicide
Maneb Tomato & Vegetable Fungicide
Tomato Potato Dust

LEUCOTHOE
Bayleton
Procide
Systemic Fungicide

LILAC
Bayleton
Bordeaux Mix
Bordeaux Mixture
Copper Fungicide
Funginex
Garden Sulphur
Kop-R-Spray
Lawn, Ornamental & Vegetable Fungicide
Lime Sulfur Solution
Microcop
Orthenex Insect & Disease Control
Procide
Systemic Fungicide
Systemic Fungicide 3336
Wettable Dusting Sulfur

LILY
Bordeaux Mix
Bordeaux Mixture
Bravo
Bulb Dust
Daconil
Daconil 2787
Dithane M-45
Kop-R-Spray
Lawn, Ornamental & Vegetable Fungicide
Mancozeb
Mancozeb Fungicide
Mancozeb Plant Fungicide
Maneb Tomato & Vegetable Fungicide
Multi-Purpose Fungicide

LINDEN
Bordeaux Mix

Bordeaux Mixture
Bordeaux Powder
Kop-R-Spray

LIPSTICK PLANT
Lawn, Ornamental & Vegetable Fungicide

LOCUST
Bayleton
Procide
Systemic Fungicide

MAGNOLIAS
Copper Fungicide
Dithane M-45
Lawn, Ornamental & Vegetable Fungicide
Liquid Copper
Mancozeb
Mancozeb Plant Fungicide

MANGOES
Copper Fungicide
Kop-R-Spray
Liquid Copper
Wettable Dusting Sulfur

MAPLE
Bayleton
Bordeaux Mix
Bordeaux Mixture
Bordeaux Powder
Copper Fungicide
Copper Spray or Dust
Dithane M-45
Garden Sulphur
Kop-R-Spray
Lawn, Ornamental & Vegetable Fungicide
Liquid Copper
Mancozeb
Orthenex Insect & Disease Control I
Procide
Systemic Fungicide
Wettable Dusting Sulfur

MARIGOLD
Bayleton
Copper Spray or Dust
Dithane M-45
Kop-R-Spray
Lawn, Ornamental & Vegetable Fungicide
Mancozeb
Mancozeb Plant Fungicide
Orthenex Insect & Disease Control I
Procide
Systemic Fungicide

MELONS
Bayleton

Bordeaux Powder
Bravo
Captan 50% WP
Copper Dragon
Copper Fungicide
Copper Spray or Dust
Daconil
Daconil 2787
Dithane M-45
Foli-Cal
Fungicide
Kop-R-Spray
Lawn & Garden Fungicide
Lawn, Ornamental & Vegetable Fungicide
Liquid Copper
Liquid Copper Fungicide
Mancozeb
Mancozeb Plant Fungicide
Maneb Tomato & Vegetable Fungicide
Multi-Purpose Fungicide
Tomato Potato Dust
Wettable Dusting Sulfur

MOCK ORANGE
Bayleton
Procide
Systemic Fungicide

MOUNTAIN LAUREL
Bayleton
Bravo
Daconil
Daconil 2787
Dithane M-45
Lawn, Ornamental & Vegetable Fungicide
Procide
Systemic Fungicide

MUSTARD
Copper Dragon
Wettable Dusting Sulfur

MYRTLE
Dormant Disease Control
Orthenex Insect & Disease Control I

NARCISSUS
Bulb Dust
Dithane M-45
Fungicide
Lawn, Ornamental & Vegetable Fungicide
Mancozeb
Mancozeb Plant Fungicide

NASTURTIUM
Kop-R-Spray

97

FUNGICIDES

NECTARINES
Bravo
Captan
Copper Fungicide
Daconil
Daconil 2787
Kop-R-Spray
Lawn, Ornamental & Vegetable Fungicide
Liquid Copper
Multi-Purpose Fungicide
Sulfur Dust
Sulfur Plant Fungicide
Systemic Fungicide
Systemic Fungicide 3336
Wettable Dusting Sulfur

NEPHTHYTIS
Bayleton
Procide
Systemic Fungicide

NINEBARK
Bayleton
Procide
Systemic Fungicide

OAK
Bayleton
Bordeaux Mix
Bordeaux Mixture
Bordeaux Powder
Bravo
Copper Fungicide
Daconil 2787
Dithane M-45
Kop-R-Spray
Lawn, Ornamental & Vegetable Fungicide
Liquid Copper
Mancozeb
Mancozeb Plant Fungicide
Multi-Purpose Fungicide
Orthenex Insect & Disease Control I
Procide

OLEANDER
Orthenex Insect & Disease Control I

OLIVES
Kop-R-Spray

ONIONS
Bravo
Copper Fungicide
Daconil 2787
Kop-R-Spray
Lawn, Ornamental & Vegetable Fungicide
Liquid Copper
Liquid Copper Fungicide
Mancozeb
Mancozeb Fungicide
Mancozeb Plant Fungicide
Maneb Tomato & Vegetable Fungicide
Multi-Purpose Fungicide
Sulfur Dust
Wettable Dusting Sulfur

ORCHIDS
Dithane M-45
Mancozeb
Mancozeb Plant Fungicide
Orthenex Insect & Disease Control I

OREGON GRAPE
Bravo
Daconil
Daconil 2787
Lawn, Ornamental & Vegetable Fungicide
Multi-Purpose Fungicide

ORNAMENTALS (general)
Copper Dragon
Dormant Disease Control
Flower Fruit & Vegetable Garden Fungicide
Lawn & Garden Fungicide
Liquid Fruit Tree Spray
Orthenex Insect & Disease Control
Polysul
Rose & Flower Spray or Dust
Rose & Ornamental Fungicide
Sulfur Dust
Sulfur Plant Fungicide
Sulf-R-Spray
Systemic Fungicide
Systemic Fungicide 3336
Wettable Dusting Sulfur

OXALIS
Funginex

OYSTER PLANT
Bravo
Daconil 2787
Lawn, Ornamental & Vegetable Fungicide
Multi-Purpose Fungicide

PACHYSANDRA
Bravo
Copper Fungicide
Dithane M-45
Lawn, Ornamental & Vegetable Fungicide
Liquid Copper
Mancozeb

PALMS
Bordeaux Mixture
Copper Fungicide
Daconil 2787
Kop-R-Spray
Lawn, Ornamental & Vegetable Fungicide
Liquid Copper
Multi-Purpose Fungicide

PANSY
Bayleton
Bordeaux Mix
Bordeaux Mixture
Dithane M-45
Kop-R-Spray
Lawn, Ornamental & Vegetable Fungicide
Mancozeb
Mancozeb Fungicide
Mancozeb Plant Fungicide
Maneb Tomato & Vegetable Fungicide
Procide
Systemic Fungicide

PAPAYA
Bravo
Daconil 2787
Dithane M-45
Lawn, Ornamental & Vegetable Fungicide
Multi-Purpose Fungicide

PARLOR PALM
Bravo
Lawn, Ornamental & Vegetable Fungicide

PARSNIP
Daconil 2787
Lawn, Ornamental & Vegetable Fungicide
Multi-Purpose Fungicide

PASSION FRUIT
Multi-Purpose Fungicide

PAULOWNIA (Empress Tree)
Bayleton
Procide
Systemic Fungicide

PEAS
Captan 50% WP
Copper Fungicide
Flower Fruit & Vegetable Garden Fungicide
Liquid Copper
Sulfur Dust
Sulfur Plant Fungicide
Wettable Dusting Sulfur

FUNGICIDES

USAGE OF PRODUCTS

PEACHES
Bordeaux Mix
Bordeaux Mixture
Bordeaux Powder
Bravo
Captan
Captan 50% WP
Captan Wettable
Complete Fruit Tree Spray
Copper Fungicide
Daconil
Daconil 2787
Dormant Disease Control
Ferbam
Fruit Spray Concentrate
Fruit Tree Spray
Garden Sulphur
Home Orchard Spray
Kop-R-Spray
Lawn, Ornamental & Vegetable Fungicide
Lime Sulfur Solution
Lime Sulfur Spray
Liqui-Cop
Liquid Copper
Liquid Fruit Tree Spray
Liquid Sulfur
Microcop
Multi-Purpose Fungicide
Polysul
Sulfur Dust
Sulfur Plant Fungicide
Sulf-R-Spray
Systemic Fungicide
Systemic Fungicide 3336
Wettable Dusting Sulfur

PEANUTS
Copper Fungicide
Lawn, Ornamental & Vegetable Fungicide
Liquid Copper

PEAR (Flowering)
Bayleton
Procide
Systemic Fungicide

PEARS
Bayleton
Bordeaux Mixture
Copper Fungicide
Copper Spray or Dust
Dormant Disease Control
Ferbam
Flower Fruit & Vegetable Garden Fungicide
Foli-Cal
Garden Sulphur
Home Orchard Spray
Kop-R-Spray
Lime Sulfer Spray
Liquid Sulfur
Microcop
Polysul
Sulfur Dust
Sulf-R-Spray
Sulfur Plant Fungicide
Wettable Dusting Sulfur

PECANS
Bordeaux Mixture
Systemic Fungicide
Systemic Fungicide 3336

PEONY
Bordeaux Mix
Bordeaux Mixture
Dithane M-45
Ferbam
Kop-R-Spray
Mancozeb
Maneb Tomato & Vegetable Fungicide

PEPERIOMIA
Dithane M-45
Lawn, Ornamental & Vegetable Fungicide
Mancozeb

PEPPERS
Copper Dragon
Copper Fungicide
Foli-Cal
Kop-R-Spray
Liquid Copper
Liquid Copper Fungicide
Tomato Potato Dust
Wettable Dusting Sulfur

PETUNIA
Bayleton
Bravo
Daconil
Daconil 2787
Lawn, Ornamental & Vegetable Fungicide
Multi-Purpose Fungicide
Orthenex Insect & Disease Control I
Procide
Systemic Fungicide

PHILODENDRON
Bravo
Copper Fungicide
Daconil 2787
Dithane M-45
Fungicide
Lawn & Garden Fungicide
Lawn, Ornamental & Vegetable Fungicide
Liquid Copper
Mancozeb
Mancozeb Plant Fungicide
Multi-Purpose Fungicide

PHLOX
Bayleton
Bordeaux
Bordeaux Mix
Copper Spray or Dust
Funginex
Garden Sulphur
Kop-R-Spray
Lawn, Ornamental & Vegetable Fungicide
Orthenex Insect & Disease Control
Procide
Systemic Fungicide
Wettable Dusting Sulfur

PHOTINIA
Bayleton
Bravo
Daconil
Daconil 2787
Disease Control
Dithane M-45
Funginex
Lawn, Ornamental & Vegetable Fungicide
Mancozeb
Multi-Purpose Fungicide
Orthenex Insect & Disease Control
Procide
Systemic Fungicide

PIERIS (Andromeda)
Bravo
Daconil 2787
Lawn, Ornamental & Vegetable Fungicide
Multi-Purpose Fungicide

PINES
Bayleton
Bordeaux Mix
Bordeaux Mixture
Bravo
Copper Fungicide
Copper Spray or Dust
Daconil 2787
Kop-R-Spray
Lime Sulfur Solution
Liquid Copper
Liquid Copper Fungicide
Multi-Purpose Spray
Procide
Systemic Fungicide

PISTACHIOS
Sulfur Dust

FUNGICIDES

PLANETREE
Daconil 2787
Lawn, Ornamental & Vegetable Fungicide
Multi-Purpose Fungicide

PLEOMELE
Dithane M-45
Mancozeb

PLUMS
Bordeaux Mixture
Bravo
Captan
Captan Wettable
Daconil
Daconil 2787
Fruit Spray Concentrate
Garden Sulpher
Kop-R-Spray
Lawn, Ornamental & Vegetable Fungicide
Lime Sulfur Spray
Liquid Sulfur
Multi-Purpose Fungicide
Sulfur Dust
Sulfur Plant Fungicide
Systemic Fungicide
Systemic Fungicide 3336
Wettable Dusting Sulfur

POINSETTIAS
Dithane M-45
Fungicide
Lawn, Ornamental & Vegetable Fungicide
Mancozeb
Orthenex Insect & Disease Control I
Rose & Ornamental Fungicide
Systemic Fungicide
Systemic Fungicide 3336

POPLAR
Bayleton
Bravo
Daconil
Daconil 2787
Funginex
Lawn, Ornamental & Vegetable Fungicide
Lime Sulfur Solution
Multi-Purpose Fungicide
Orthenex Insect & Disease Control I
Procide
Systemic Fungicide

POTATOS
Bordeaux Powder
Borfdeaux Mixture
Bravo
Copper Dragon
Copper Fungicide
Copper Spray or Dust
Daconil
Daconil 2787
Dithane M-45
Flower Fruit & Vegetable Garden Fungicide
Foli-Cal
Kop-R-Spray
Lawn, Ornamental & Vegetable Fungicide
Liquid Copper
Liquid Sulfur
Mancozeb
Mancozeb Fungicide
Mancozeb Plant Fungicide
Maneb Tomato & Vegetable Fungicide
Microcop
Multi-Purpose Fungicide
Tomato Dust
Tomato Potato Dust

POTATO (Seed Piece Treatment)
Captan

POTENTILLA
Bayleton
Procide
Systemic Fungicide

PRAYER PLANT (Maranta)
Bravo
Daconil 2787
Lawn, Ornamental & Vegetable Fungicide
Multi-Purpose Fungicide

PRIVET (Ligustrum)
Bayleton
Bravo
Daconil 2787
Dithane M-45
Garden Sulphur
Moncozeb
Multi-Purpose Fungicide
Procide
Rose & Ornamental Spray
Systemic Fungicide
Systemic Fungicide 3336
Wettable Dusting Sulfur

PRUNES
Bravo
Captan
Captan Wettable
Daconil
Daconil 2787
Garden Sulphur
Kop-R-Spray
Lawn, Ornamental & Vegetable Fungicide
Liquid Sulfur
Sulfur Dust
Sulfur Plant Fungicide
Systemic Fungicide
Systemic Fungicide 3336
Wettable Dusting Sulfur

PUMPKINS
Bayleton
Bravo
Copper Dragon
Copper Fungicide
Copper Spray or Dust
Daconil
Daconil 2787
Lawn & Garden Fungicide
Lawn, Ornamental & Vegetable Fungicide
Liquid Copper
Liquid Copper Fungicide
Maneb Tomato & Vegetable Fungicide
Multi-Purpose Fungicide
Tomato Potato Dust

PURPLELEAF SAND CHERRY
Daconil 2787
Multi-Purpose Fungicide

PURPLE PASSION
Orthenex Insect & Disease Control I

PYRACANTHA (Firethorn)
Bayleton
Bravo
Copper Fungicide
Dithane M-45
Lawn, Ornamental & Vegetable Fungicide
Liquid Copper
Mancozeb
Mancozeb Plant Fungicide
Microcop
Orthenex Insect & Disease Control I
Procide
Rose & Ornamental Fungicide
Systemic Fungicide
Systemic Fungicide 3336

QUINCE (Flowering)
Bravo
Daconil
Daconil 2787
Lawn, Ornamental & Vegetable Fungicide
Multi-Purpose Fungicide

REDWOOD
Lawn, Ornamental & Vegetable Fungicide

FUNGICIDES

RHODODENDRON
Bayleton
Bordeaux Mix
Bordeaux Mixture
Bravo
Copper Fungicide
Copper Spray or Dust
Daconil
Daconil 2787
Disease Control
Dithane M-45
Funginex
Garden Sulphur
Kop-R-Spray
Lawn, Ornamental & Vegetable
    Fungicide
Mancozeb
Microcop
Multi-Purpose Fungicide
Procide
Rose & Ornamental Fungicide
Sulfur Dust
Systemic Fungicide
Systemic Fungicide 3336
Thiomyl
Wettable Dusting Sulfur

ROSES
Bayleton
Bravo
Captan
Captan 50% WP
Captan Wettable
Complete Fruit Tree Spray
Copper Dragon
Copper Fungicide
Copper Spray or Dust
Daconil
Daconil 2787
Disease Control
Dithane M-45
Dormant Disease Control
Ferbam
Flower Fruit & Vegetable Garden
    Fungicide
Fruit Tree Spray
Fungicide
Garden Sulphur
Kop-R-Spray
Lawn & Garden Fungicide
Lawn, Ornamental & Vegetable
    Fungicide
Lime Sulfur Solution
Liquid Copper
Liquid Copper Spray
Liquid Fruit Tree Spray
Mancozeb
Mancozeb Fungicide
Mancozeb Plant Fungicide
Maneb Tomato & Vegetable
    Fungicide
Microcop
Multi-Purpose Fungicide
Orthenex Insect & Disease Control
Orthenex Insect & Disease Control I
Polysul
Procide
Rose & Flower Spray or Dust
Rose & Flower Insect & Disease
    Concentrate
Rose & Ornamental Fungicide
Sulf-R-Spray
Sulfer Dust
Sulfur Plant Fungicide
Systemic Fungicide
Systemic Fungicide 3336
Wettable Dusting Sulfur

RUSSIAN OLIVE
Bayleton
Procide
Systemic Fungicide

RUTABEGAS
Sulfur Dust
Wettable Dusting Sulfur

SALVIA
Bayleton
Orthenex Insect & Disease Control I
Procide
Systemic Fungicide

SAND CHERRY
Bravo
Daconil
Lawn, Ornamental & Vegetable
    Fungicide

SCHEFFLERA
Dithane M-45
Mancozeb

SEDUM
Bayleton
Procide
Systemic Fungicide

SHADE TREES (general)
Lime Sulfur Solution
Lime Sulfur Spray
Rose & Ornamental Fungicide
Systemic Fungicide 3336

SHALLOTS
Bravo
Daconil 2787
Lawn, Ornamental & Vegetable
    Fungicide
Multi-Purpose Fungicide

SHRUBS (general)
Copper Fungicide
Copper Spray or Dust
Dormant Disease Control
Fung-Away
Lime Sulfur Solution
Lime Sulfur Spray
Microcop
Polysul
Sulfur Dust
Sulfur Plant Fungicide
Sulf-R-Spray
Systemic Fungicide 3336
Thiomyl

SNAPDRAGON
Bayleton
Dithane M-45
Ferbam
Fungicide
Funginex
Garden Sulphur
Kop-R-Spray
Mancozeb
Maneb Tomato & Vegetable
    Fungicide
Orthenex Insect & Disease Control
Orthenex Insect & Disease Control I
Procide
Sulfur Dust
Systemic Fungicide
Wettable Dusting Sulfur

SPINACH
Copper 50%WP
Copper Dragon
Kop-R-Spray

SPIRAEA
Bayleton
Lawn, Ornamental & Vegetable
    Fungicide
Orthenex Insect & Disease Control I
Procide
Systemic Fungicide

SPRUCE
Bordeaux Mixture
Daconil 2787
Kop-R-Spray

SQUASH
Bayleton
Bordeaux Powder
Bravo
Captan 50% WP
Copper Dragon
Copper Fungicide
Copper Spray or Dust
Daconil
Daconil 2787
Dithane M-45
Flower Fruit & Vegetable Garden
    Fungicide

USAGE OF PRODUCTS

FUNGICIDES

Kop-R-Spray
Lawn & Garden Fungicide
Lawn, Ornamental & Vegetable Fungicide
Liquid Copper
Liquid Copper Fungicide
Mancozeb
Maneb Tomato & Vegetable Fungicide
Multi-Purpose Fungicide
Tomato Potato Dust
Wettable Dusting Sulfur

STATICE

Bravo
Daconil 2787
Dithane M-45
Lawn, Ornamental & Vegetable Fungicide
Mancozeb
Multi-Purpose Fungicide

STOCKS

Copper Spray or Dust
Kop-R-Spray

STRAWBERRIES

Bordeaux Mix
Bordeaux Mixture
Captan
Captan 50% WP
Captan Wettable
Complete Fruit Tree Spray
Copper Dragon
Copper Fungicide
Copper Spray or Dust
Flower Fruit & Vegetable Garden Fungicide
Fruit Spray Concentrate
Fruit Tree Spray
Home Orchard Spray
Kop-R-Spray
Liquid Copper
Liquid Fruit Tree Spray
Liquid Sulfur
Sulfur Dust
Sulfur Plant Fungicide
Wettable Dusting Sulfur

SUMAC

Dithane M-45
Mancozeb

SUNFLOWER

Bayleton
Procide
Systemic Fungicide

SWEET PEAS

Bayleton
Dormant Disease Control
Garden Sulphur
Kop-R-Spray

SWISS CHARD

Captan 50% WP

SYCAMORE (Plane Tree)

Bayleton
Bordeaux Mix
Bordeaux Mixture
Bravo
Copper Fungicide
Daconil
Daconil 2787
Funginex
Kop-R-Spray
Lawn, Ornamental & Vegetable Fungicide
Liquid Copper
Liquid Copper Spray
Multi-Purpose Fungicide
Orthenex Insect & Disease Control I
Procide
Systemic Fungicide

SYNGONIUM (Nephthytis)

Bayleton
Bravo
Daconil 2787
Dithane M-45
Lawn, Ornamental & Vegetable Fungicide
Mancozeb
Multi-Purpose Fungicide
Procide

TOMATOS

Bordeaux Mix
Bordeaux Mixture
Bordeaux Powder
Bravo
Copper Dragon
Copper Fungicide
Copper Spray or Dust
Daconil
Daconil 2787
Dithane M-45
Foli-Cal
Fungicide
Kop-R-Spray
Lawn & Garden Fungicide
Lawn, Ornamental & Vegetable Fungicide
Liquid Copper
Liquid Copper Fungicide
Liquid Sulfur
Mancozeb
Mancozeb Fungicide
Mancozeb Plant Fungicide
Maneb Tomato & Vegetable Fungicide
Microcop
Multi-Purpose Fungicide
Sulfur Dust
Tomato Blossom End Rot Spray
Tomato Dust
Tomato Potato Dust
Wettable Dusting Sulfur

TREES (general)

Dormant Disease Control
Foli-Cal
Fung-Away
Sulfur Dust
Systemic Fungicide
Systemic Fungicide 3336
Thiomyl

TULIPS

Bordeaux Mix
Bordeaux Mixture
Bulb Dust
Dithane M-45
Ferbam
Fungicide
Kop-R-Spray
Lawn, Ornamental & Vegetable Fungicide
Mancozeb
Mancozeb Fungicide
Mancozeb Plant Fungicide
Rose & Ornamental Fungicide
Systemic Fungicide 3336

TULIP TREE

Bordeaux Mixture
Kop-R-Spray

TURF

Bayleton
Bravo
Captan
Captan 50% WP
Captan Wettable
Daconil
Daconil 2787
Disease Control
Dithane M-45
Foli-Cal
Fung-Away
Fungicide
Lawn & Garden Fungicide
Lawn, Ornamental & Vegetable Fungicide
Mancozeb Fungicide
Maneb Tomato & Vegetable Fungicide
Multi-Purpose Fungicide
Procide G
Systemic Fungicide
Systemic Fungicide 3336
Thiomyl
Turf Fungicide

FUNGICIDES

TURF (seed beds)
Captan

TURNIPS
Copper Dragon
Sulfur Dust
Wettable Dusting Sulfur

VEGETABLES (general)
Foli-Cal

VEGETABLE SEEDLINGS
Copper Fungicide
Microcop

VENUS FLYTRAP
Dithane M-45
Mancozeb

VIBURNUM
Bayleton
Bravo
Daconil
Daconil 2787
Dithane M-45
Lawn, Ornamental & Vegetable Fungicide
Mancozeb
Multi-Purpose Fungicide
Orthenex Insect & Disease Control I
Procide
Systemic Fungicide
Systemic Fungicide 3336

VINES (general)
Foli-Cal

VIOLETS
Kop-R-Spray

VIRGINIA CREEPER
Kop-R-Spray

VITEX (Chaste Tree)
Bayleton
Procide
Systemic Fungicide

WALNUT
Bayleton
Copper Fungicide
Kop-R-Spray
Lawn, Ornamental & Vegetable Fungicide
Liqui-Cop
Liquid Copper
Mancozeb
Procide
Systemic Fungicide
Wettable Dusting Sulfur

WATERMELON
Bayleton
Bordeaux Powder
Bravo
Copper Fungicide
Daconil
Daconil 2787
Dithane M-45
Lawn & Garden Fungicide
Lawn, Ornamental & Vegetable Fungicide
Liquid Copper
Liquid Copper Fungicide
Mancozeb
Maneb Tomato & Vegetable Fungicide
Multi-Purpose Fungicide
Tomato Potato Dust

WILLOW
Bayleton
Bordeaux Mix
Bordeaux Mixture
Kop-R-Spray
Lime Sulfur Solution
Orthenex Insect & Disease Control I
Procide
Systemic Fungicide

YAUPON
Orthenex Insect & Disease Control I

YEW
Bordeaux Mix
Bordeaux Mixture
Kop-R-Spray

ZEBRA PLANT
Lawn, Ornamental & Vegetable Fungicide

ZINNIA
Bayleton
Bravo
Daconil 2787
Dithane M-45
Ferbam
Fungicide
Funginex
Garden Sulphur
Kop-R-Spray
Lawn & Garden Fungicide
Lawn, Ornamental & Vegetable Fungicide
Mancozeb
Mancozeb Plant Fungicide
Maneb Tomato & Vegetable Fungicide
Multi-Purpose Fungicide
Orthenex Insect & Disease Control
Orthenex Insect & Disease Control I
Procide
Systemic Fungicide
Wettable Dusting Sulfur

A Guide To
Lawn, Garden & Home
Pest Control Products

Section IV   **Plant Names**

        Scientific Names   106
        (Common Name Cross Reference)

        Common Names   114
        (Scientific Name Cross Reference

SCIENTIFIC NAME – *Common Name Cross Reference*

Abeliophyllum	Forsythia (white)	*Anemone*	Pasque Flower
Abelmoschus	Musk Mallow	*Anemone*	Wind Flower
Abies	Fir	*Anthemis*	Marguerite (Golden)
Abronia	Sand Verbena	*Antigonon*	Coral Vine
Abutilon	Chinese Bell Flower	*Antigonon*	Mexican Creeper
Abutilon	Flowering Maple	*Antigonon*	Queens Wreath
Acacia	Purple Leaf	*Antirrhimum*	Snapdragon
Acacia	Wattle	*Aphelandra*	Zebra Plant
Acalypha	Beefsteak Plant	*Aquilegia*	Columbine
Acalypha	Chenille Plant	*Arabis*	Rock Cress
Acalypha	Copper Plant	*Arachniodes*	Holly Fern
Acalypha	Copperleaf	*Aralia*	Japanese Angelica Tree
Acalypha	Firetails	*Aralia*	Walking Stick
Acanthus	Bear's Breech	*Araucaria*	Monkey Puzzle Tree
Acer	Box Elder	*Araucaria*	Norfolk Island Pine
Acer	Maple	*Arbutus*	Madrone
Achillea	Milfoil	*Arbutus*	Strawberry Tree
Achillea	Yarrow	*Arctostaphylos*	Bearberry
Achimenes	Star of India	*Arctostaphylos*	Manzanita
Aconitum	Monkshood	*Arctotheca*	Cape Weed
Acorus	Sweet Grass (Flag)	*Arecastrum*	Queen Palm
Actaea	Baneberry	*Arenaria*	Moss Sandwort
Actinidia	Chinese Gooseberry	*Arisaemia*	Cobra Lily
Actinidia	Kiwi	*Arisaemia*	Jack in the Pulpit
Adenium	Desert Rose	*Armeria*	Sea Pinks
Adenophora	Lady Bells	*Armeria*	Thrift
Adiantum	Maidenhead Fern	*Aronia*	Chokeberry
Aegopodium	Bishop's Weed	*Artemisia*	White Sage
Aeschynanthus	Lipstick Plant	*Artemisia*	Wormwood
Aesculus	Buckeye	*Aruncus*	Goat's Beard
Aesculus	Horse Chestnut	*Asarum*	European Ginger
Agapanthus	African Lily	*Asarum*	Ginger
Agapanthus	Lily of the Nile	*Asclepias*	Blood Flower
Agastiache	Hyssop	*Asclepias*	Butterfly Weed
Agave	Century Plant	*Asclepias*	Milkweed
Ageratum	Floss Flower	*Asimina*	Paw Paw
Aglaonema	Chinese Evergreen	*Asparagus*	Sprengeri Fern
Ajuga	Bugle Flower	*Aspidistra*	Cast-Iron Plant
Ajuga	Bugleweed	*Aster*	Michaelmas Daisy
Ajuga	Carpet Bugle	*Astibe*	False Spirea
Albizia	Mimosa	*Astibe*	Goat's Beard
Albizia	Silktree	*Astrantia*	Masterwort
Alcea	Hollyhock	*Asystasia*	Coromandel
Alchemilla	Lady's Mantle	*Athyrium*	Lady Fern
Alnus	Alder	*Athyrium*	Speenwort Fern
Alocasia	Elephant's Ear	*Aubrieta*	Rock Cress
Aloysia	Lemon Verbena	*Aucuba*	Gold Dust Plant
Alstroemeria	Peruvian Lily	*Aurinia*	Basket of Gold
Alternanthera	Calico Plant	*Baccharis*	Coyote Bush
Alternanthera	Joseph's	*Bambusa*	Bamboo
Althea	Rose of Sharon	*Bapitsia*	Indigo
Alyogyne	Lilac Hibiscus	*Beaucarnea*	Ponytail Palm
Alyssoides	Bladder Pod	*Belamcanda*	Blackberry Lily
Amalanchier	Juneberry	*Bellis*	English Daisy
Amalanchier	Shadbush	*Berberis*	Barberry
Amelanchier	Serviceberry	*Berberis*	Crimson Pygmy
Amorpha	False Indigo	*Betula*	Birch
Ampelopsis	Porcelain Vine	*Bignonia*	Trumpet Crossvine
Amsonia	Blue Star	*Bletilla*	Chinese Ground Orchid
Anacyclus	Mt. Atlas Daisy	*Bougainvillea*	Raspberry Ice
Anagallis	Pimpernel	*Brachychiton*	Bottle Tree
Andromeda	Bog Rosemary	*Brachycome*	Swan River Daisy

Common Name Cross Reference – SCIENTIFIC NAME

Brassaia	Schefflera	*Centranthus*	Jupiter's Beard
Brassaia	Umbrella Tree	*Centranthus*	Keys of Heaven
Brassicaceae	Flowering Kale	*Centranthus*	Valerian
Browallia	Amethyst Flower	*Cephalanthus*	Buttonbush
Brugmansia	Angel's Trumpet	*Cephalotaxus*	Plum Yew
Brunnera	Forget-Me-Not	*Cerastium*	Snow in Summer
Buddleia	Butterfly Bush	*Ceratonia*	Carob
Butia	Pindo Palm	*Ceratostigma*	Dwarf Plumbago
Buxus	Boxwood	*Ceratostigma*	Plumbago
Caesalpinia (Gilliesii)	Bird of Paradise (Bush)	*Cercidiphyllum*	Katsura Tree
Calathea	Peacock Plant	*Cercidium*	Palo Verde
Calendula	Pot Marigold	*Cercis*	Redbud
Callicarpa	Beauty Berry	*Cercocarpus*	Mountain Mahogany
Callirhoe	Poppy Mallow	*Cestrum*	Night-blooming Jasmine
Callistemon	Bottlebrush	*Chaenomeles*	Quince
Callistephus	Aster (Stokes)	*Chamaecyparis*	Cedar
Calluna	Heather	*Chamaecyparis*	False Cypress
Calocedrus	Incense Cedar	*Chamaecyparis*	Hinoki Cypress
Calonyction	Moon Flower	*Chamaedorea*	Bamboo Palm
Caltha	Marsh Marigold	*Chamaedorea*	Palm
Calycanthus	Carolina Allspice	*Chamaedorea*	Parlor Palm
Camassia	Camass	*Chamaerop*	Fan Palm
Campanula	Bell Flower	*Cheiranthus*	Wall Flower
Campanula	Canterberry Bells	*Cheiranthus*	Wallflower
Campanula	Harebell	*Cheirodendron*	Lapa Lapa
Campsis	Trumpet Creeper	*Chelone*	Snakehead
Caprosma	Mirror Plant	*Chelone*	Turtlehead
Capsicum	Pepper	*Chilopsis*	Desert Willow
Caragana	Peashrub	*Chimonamthus*	Wintersweet
Caragana	Siberian Pea Tree	*Chionanthus*	Fringe Tree
Carcia	Papaya	*Chiorophytum*	Spider Plant
Cardiospermum	Balloon Vine	*Choisya*	Mexican Orange
Cardiospermum	Heartpea	*Chrysalidocarpus*	Areca Palm
Carex	Sedge	*Chrysanthemum*	Cushion Mums
Carissa	Natal Plum	*Chrysanthemum*	Feverfew
Carpinus	Hornbean	*Chrysanthemum*	Marguerite
Carpobrotus (perennials)	Iceplant	*Chrysanthemum*	Ox-Eye Daisy
Carya	Hickory	*Chrysanthemum*	Painted Daisy
Carya	Pecan	*Chrysanthemum*	Shasta Daisy
Caryopteris	Blue Mist	*Chrysogonum*	Goldenstar
Caryopteris	Bluebeard	*Chrysogonum*	Green Gold
Caryota	Firecracker Palm	*Chrysopsis*	Golden Aster
Caryota	Fishtail Palm	*Cibotium*	Hawaiian Tree Fern
Cassiope	Heather	*Cimicifuga*	Bugbane
Castanea	Chestnut	*Cimicifuga*	Snake Root
Catananche	Cupid's Dart	*Cissus*	Grape Ivy
Catharanthus	Periwinkle (madagascar)	*Cissus*	Kangaroo Ivy
Catharanthus	Periwinkle (Vinca)	*Cissus*	Leaf Ivy
Ceanothus	Caramel Creeper	*Cistus*	Rock Rose
Ceanothus	Redroot Jersey Tea	*Cladrastis*	Yellow Wood
Ceanothus	Squaw Carpet	*Cleome*	Spider Plant
Ceanothus	Wild Lilac	*Clerodendrum*	Bleeding Heart Glorybower
Cedrus	Cedar	*Clethra*	Summersweet
Cedrus	Cedar of Lebanon	*Clethra*	Sweet Pepperbush
Celastrus	Bittersweet	*Clivia*	Kafer Lily
Celosia	Cockscomb	*Codiaeum*	Croton
Celtis	Hackberry	*Coffea*	Coffee Tree
Ceniza	Texas Sage	*Coix*	Job's Tears
Centaurea	Bachelor Buttons	*Colocasia*	Elephant Ears
Centaurea	Dusty Miller	*Consolida*	Larkspur
Centaurea	Persian Cornflower	*Convallaria*	Lily of the Valley
		Cordyline	Cracena

SCIENTIFIC NAME – *Common Name Cross Reference*

Cordyline	Hawaiian Ti	*Dodecatheon*	Shooting Star
Coreopsis	Golden Daisy	*Dodonaea*	Purple Hopseed Bush
Coreopsis	Tickseed	*Doronicum*	Leopards Bane
Cornus	Bunchberry	*Dracaena*	Corn Plant
Cornus	Cornelian Cherry	*Dracaena*	Marginata
Cornus	Dogwood	*Dracaena*	Reflexa
Coronilla	Crownvetch	*Dracquephalium*	Dragonhead
Correa	Fuchsia (Australian)	*Drosanthemum*	Iceplant (trailing)
Cortaderia	Pampas Grass	*Dryopteris*	Wood Fern
Corydalis	Bleeding Heart	*Duranta*	Sky Flower
Corylopsis	Winterhazel	*Dyssodia*	Dahlberg Daisy
Corylus	Filbert	*Echinacea*	Coneflower
Corylus	Hazelnut	*Echinacea*	Purple Coneflower
Cotinus	Purple Fringe	*Echinops*	Globe Thistle
Cotinus	Smokebush	*Elaeagnus*	Autumn Olive
Cotinus	Smoketree	*Elaeagnus*	Russian Olive
Cotoneaster	Bearberry	*Elaeagnus*	Silverberry
Cotoneaster	Coral Beauty	*Endymion*	Spanish Bluebell
Cotoneaster	Cranberry	*Epimedium*	Barrenwort
Crassula	Jade Plant (Tree)	*Epipremnum*	Pothos Ivy
Crataegus	Hawthorn	*Eremurus*	Foxtail Lily
Cryophytum (Annual)	Iceplant	*Eribobotrya*	Loquat
Cryptomeria	Japanese Cedar	*Erica*	Heath
Cryptostesia	Rubber Vine	*Erica*	Heather
Cupaniopsis	Carrot Wood	*Erigeron*	Daisy Fleabane
Cuphea	False Heather	*Eriophyllum*	Woody Sunflower
Cuphea	Mexican Heather	*Erymgium*	Sea Holly
Cupressocyparis	Leyland Cypress	*Erysimum*	Wallflower
Cupressus	Arizona Cypress	*Erythronium*	Dogtooth Violet
Cupressus	Italian Cypress	*Eschscholzia*	California Poppy
Curcuma	Hidden Lily	*Eucalyptus*	Red Gum
Cycas	Sago Palm	*Eucalyptus*	Red Iron Bark
Cyclamen	Persian Violet	*Eucharis*	Amazon Lily
Cyperus	Umbrella Plant	*Eucomis*	Pineapple Lily
Cyrtomium	Holly Fern	*Eugenia*	Cherry Brush
Cytisus	Broom	*Eunoymus*	Winter Creeper
Cytisus	Scotchbroom	*Eunoymus*	Burning Bush
Daboecia	Heath (Irish)	*Euonymus*	Silver King
Darmera	Umbrella Plant	*Euonymus*	Spindle Tree
Datura	Angel's Trumpet	*Euonymus*	Winter Creeper
Datura	Thorn Apple	*Eupatorium*	Joe Pye Weed
Daucus	Queen Anne's Lace	*Eupatorium*	Mist Flower
Davidia	Dove Tree	*Euphorbia*	Spurge
Delosperma	Iceplant(white)(hardy)	*Eustoma*	Texas Bluebell
Delphinum	Larkspur	*Evolvulus*	Blue Daze
Dendranthema	Chrysanthemum	*Exochorda*	Pearl Bush
Dendrobium	Orchid	*Fagus*	Beech
Dennstaedtia	Hay Sentel Fern	*Fargesia*	Clump Bamboo
Dianthus	Carnation	*Fatshedera*	Botanical Wonder
Dianthus	Pink	*Fatsia*	Aralia
Dianthus	Sweet William	*Feijoa*	Guava
Diascia	Twinspur	*Felicia*	Blue Bush Daisy
Dicentra	Bleeding Heart	*Felicia*	Daisy Bush
Dicentra	Dutchman's Breeches	*Festuca*	Fescue
Dictamnus	Gas Plant	*Ficus*	Banyan
Dieffenbachia	Bali High	*Ficus*	Cuban Laurel
Dieffenbachia	Dumb Cane	*Ficus*	Fig
Diervilla	Bush Honeysuckle	*Ficus*	Indian Laurel
Digitalis	Foxglove	*Ficus*	Rubber Plant (tree)
Dimorphotheca	Cape Marigold	*Ficus*	Weeping Fig
Diospyros	Persimmon	*Filipendula*	Meadow Sweet
Dizgsotheca	False Aralia	*Filipendula*	Queen of the Prairie

Common Name Cross Reference – SCIENTIFIC NAME

Fittonia	Nerve Plant	Homeria	Cape Iris
Forsythia	Golden Bells	Hosta	Plantain Lily
Fragaria	Strawberry	Howea	Kentia Palm
Fraxinus	Ash	Hoya	Wax Plant
Gaillardia	Blanket Flower	Hylatelephium	October Plant Perilla
Gaillardia	Fiesta Daisy	Hypericum	Aaron's Beard
Galium	Sweet Woodruff	Hypericum	St. Johns Wort
Gallardia	Fiesta Daisy	Hypoestes	Polka Dot Plant
Gallardia	Indian Blanket	Iberis	Candytuft
Ganzania	Treasure Flower	Ilex	Holly
Gardenia	Cape Jasmine	Ilex	Inkberry
Garrya	Silk Tassel	Ilex	Winterberry
Gaultheria	Wintergreen	Ilex	Yaupon
Gelsemium	Caroline Jasmine	Illicium	Spicebush
Genista	Broom	Impomea	Morninglory
Genista	Woadwaxen	Incarvillae	Garden Gloxinia
Geranium	Cranesbill	Incarvillea	Hardy Gloxinia
Gerbera	Transvaal Daisy	Ipiteion	Star Flower
Geum	Avens	Isatis	Dyers Wood
Ginkgo	Maidenhair Tree	Ismene	Peruvian Daffodil
Glechoma	Ground Ivy	Itea	Sweetspire
Gleditsia	Honey Locust	Ixia	African Corn Lily
Globba	Dancing Girls	Ixia	Corn Lily
Gloriosa	Climbing Lily	Ixia	Cornbells
Gomphrena	Bachelor Buttons	Ixora	Flame of the Woods
Gomphrena	Globe Amaranth	Jasminum	Angel Wing
Grewia	Star Plant	Jasminum	Primrose Jasmine
Gymnocladus	Kentucky Coffee Tree	Juglans	Walnut
Gynura	Passion Vine	Juniperus	Cedar (Red)
Gypsophila	Baby's Breath	Juniperus	Juniper
Habenaria	Egret Lily	Justicia	Mexican Honeysuckle
Haemanthus	Blood Lily	Justicia	Shrimp Plant
Halesia	Silverbells	Kalanchoe	Panda Plant
Hamamelis	Witch Hazel	Kalmia	Mountain Laurel
Hamelia	Firebush	Kerria	Globe Flower
Hebe	Veronica	Kniphofia	Poker Plant
Hedera	Algerian Ivy	Koelreuteria	Goldenrain Tree
Hedera	English Ivy	Kolkwitzia	Beauty Bush
Hedychium	Ginger	Labumum	Golden Chain
Helianthemum	Rock Rose	Lagerstroemia	Crape Myrtle
Helianthemum	Sun Rose	Lamiastrum	Silver Frost
Helianthus	Sunflower	Lamium	Dead Nettle
Helichrysan	Straw Flower	Lamium	Yellow Archangel
Heliopsis	False Sunflower	Laptospermum	Tea Tree
Helleborus	Christmas Rose	Larix	Larch
Helleborus	Hellebore	Lathyrus	Sweet Pea
Helleborus	Lenten Rose	Laurentia	Blue Star Creeper
Hemerocallis	Day Lily	Laurus	Bay Laurel (Sweet Bay)
Hemi	Metallic Plant	Lavatera	Tree Mallow
Hepatica	European Liver-Leaf	Lavendula	Lavendar
Herniaria	Rupturewort	Leea	West Indian Holly
Hesperaloe	Yucca	Leontopodium	Edelweiss
Heuchera	Alum Root	Leonurus	Motherwort
Heuchera	Coral Bells	Leptospermum	Broom Tee Tree
Hibbertia	Guinea Gold Vine	Leptospermum	Toyon
Hibiscus	Althaea	Leucanthemum	Ox-Eye Daisy
Hibiscus	Rose Mallow	Leucanthemum	Shasta Daisy
Hibiscus	Rose of Sharon	Leucojum	Snowflake
Hippeastrum	Amaryllis	Leucophyllum	Texas Sage
Hippophae	Sea Bushthorn	Liatris	Blazing Star
Holboellia	China Blue Vine	Liatris	Gayfeather
Holodiscus	Ocean Spray	Libocedrus	Incense Cedar

SCIENTIFIC NAME – *Common Name Cross Reference*

Ligustrum	Privet	*Morus*	Mulberry
Lilium	Tiger Lily	*Murraya*	Orange Jessamine
Lillum	Lily	*Musa*	Banana
Limonium	Sea Lavendar	*Muscari*	Grape Hyacinth
Limonium	Statice	*Myosotis*	Forget-Me-Not
Lindera	Spicebush	*Myrica*	Bayberry
Linnaea	Twin Flower	*Myrica*	Wax Myrtle
Linodendron	Tulip Tree	*Myrtus*	Myrtle
Linum	Flax	*Nandina*	Heavenly Bamboo
Liquidambar	Sweetgum	*Narcissus*	Daffodil
Liriodendron	Yellow Poplar	*Nelumbo*	Lotus
Liriope	Lilyturf	*Nemopanthus*	Mountain Holly
Livistona	Chinese Fan Palm	*Neodypsis*	Triangle Palm
Lobelia	Cardinal Flower	*Nepeta*	Catnip
Lobularia	Alyssum	*Nephrolepis*	Bird's Nest Fern
Longiflorum	Lily (Easter)	*Nephrolepis*	Boston Fern
Lonicera	Honeysuckle	*Nerium*	Oleander
Lonicera	Woodbine	*Nicandra*	Shoo-Fly Plant
Loropetalum	Fringe Flower	*Nicotiana*	Flowering Tobacco
Lotus	Parrot's Beak	*Nierembergia*	Cupflower
Lunaria	Honesty Plant	*Nigella*	Love in a Mist
Lunaria	Money Plant	*Nymphaes*	Water Lily
Lupinus	Blue Bonnet	*Nyssa*	Blackgum
Lychnis	Maltese Cross	*Nyssa*	Sour Gum
Lychnis	Rose Champion	*Nyssa*	Tupelo
Lycoris	Magic Lily	*Ochna*	Mickey Mouse Bush
Lycoris	Spider Lily	*Oemleria*	Indian Plum
Lysimachia	Creeping Jenny	*Oenothera*	Evening Primrose
Lysimachia	Golden Globe	*Oenothera*	Sundrops
Lysimachia	Moneywort	*Olea*	Olive
Lythrum	Loosestrife	*Omithogalum*	Star Flower
Macleaya	Plume Poppy	*Omithogalum*	Star of Bethlehem
Maclura	Osage Orange	*Onoclea*	Sensitive Fern
Mahonia	Oregon Grape	*Ophiopogon*	Mondo Grass
Maianthemum	False Lilly of theValley	*Osmanthus*	Holly Olive
Malus	Apple	*Osmunda*	Cinnamon Fern
Malus	Crabapple	*Osmunda*	Royal Fern
Malva	Mallow	*Osteospermum*	African Daisy
Malvauiscus	Turk's Cap	*Ostrya*	Hophornbean
Mandevilla	Chilean Jasmine	*Oxalis*	Wood Sorrel
Maranta	Prayer Plant	*Oxydendrum*	Sourwood
Marshaccia	Barbara's Buttons	*Pachysandra*	Japanese Spurge
Matricaria	Chamomile	*Paeonia*	Peony
Matricaria	Feverfew	*Pancratium*	Sea Daffodil
Matthiola	Stock	*Pandanus*	Screw Pine
Meconopsis	Himalayan Blue Poppy	*Pandorea*	Bower Vine
Melaleuca	Cajeput Tree	*Pandorea*	Pandora Vine
Mella	China-Berry	*Pandorea*	Trumpet Vine
Mentha	Peppermint	*Papaver*	Poppy
Menziesia	Fool's Huckleberry	*Pardancandra*	Candy Lily
Mesembryanthemum	Ice Plant	*Parkensonia*	Palo Verde
Mesembryanthemum	Livingstone Daisy	*Parthenocissus*	Boston Ivy
Metasequoia	Redwood	*Parthenocissus*	Virginia Creeper
Mimosa	Sensitive Plant	*Passiflora*	Passionflower
Mimulus	Monkey Flower	*Paulownia*	Empress Tree
Mirabilis	Four-O-Clock	*Pavonia*	Rock Rose
Miscanthus	Eulalia Grass	*Paxistima*	Dwarf Mountain Lover
Miscanthus	Zebra Grass	*Paxistima*	Oregon Boxwood
Moluccella	Bells of Ireland	*Pelargonium*	Geranium
Monarda	Bee Balm	*Pennisetum*	Fountain Grass
Monstera	Philodendron	*Pennisetum*	Red Fountain Grass
Morea	Fortnight Lily	*Penstemon*	Beard Tongue

Common Name Cross Reference – SCIENTIFIC NAME

Perilla	Beefsteak Plant	*Prunus*	Laurel Cherry
Perovskia	Russian Sage	*Prunus*	Peach
Persea	Avocado	*Prunus*	Plum
Persea	Red Bay	*Prunus*	Purple Leaf Sand Cherry
Petrorhagia	Tunic Flower	*Pseuderanthemum*	Shooting Star
Phalaris	Gardeners Garters	*Pseudocydonia*	Chinese Quince
Phaseolius	Giant Snail Vine	*Pseudolarix*	Golden Larch
Phellodendron	Amur Corktree	*Pseudotsuga*	Douglas Fir
Philadelphus	Mock Orange	*Ptelea*	Hoptree
Philodendron	Selloum	*Pteretis*	Ostrich Fern
Phlox	Moss Pink	*Pteretis*	Table Fern
Phlox	Thrift	*Pteridium*	Bracken Fern
Phoenix	Date Palm	*Pulmonania*	Lungwort
Phormium	New Zealand Flax	*Pulsatilla*	Pasque Flower
Physocarpus	Ninebark	*Punica*	Pomegranate
Physostegia	Dragonhead	*Pyracantha*	Firethorn
Physostegia	False Dragonhead	*Pyrus*	Mountain Ash
Physostegia	Obedient Plant	*Pyrus*	Pear
Picea	Spruce	*Quercus*	Oak
Pieris	Japanese Andromeda	*Ranunculus*	Buttercup
Pieris	LilyoftheValley(shrub)	*Raphioepis*	Indian Hawthorn
Pilea	Creeping Charlie	*Raphiolepis*	Pink Lady
Pinus	Pine	*Ratibida*	Coneflower
Pistacia	Chinese Pistache	*Ravenea*	Majestic Palm
Pittsporium	Robira	*Rhamnus*	Buckthorn
Planera	Water Elm	*Rhamnus*	Coffeeberry
Platanus	Plane Tree	*Rhamnus*	Tallhedge
Platanus	Sycamore	*Rhapis*	Lady Palm
Platycerium	Staghorn Fern	*Rhexia*	Meadow Beauty
Platycladus	Oriental Arborvitae	*Rhodedendron*	Azalea
Platycodon	Balloon Flower	*Rhoeo*	Oyster Plant
Plectranthus	Spur Flower	*Rhus*	Sumac
Plectranthus	Swedish Ivy	*Ribes*	Alpine Currant
Pleioblastuis	Bamboo (dwarf)	*Robinia*	Locust (Black)
Pleione	Terrestrial Orchid	*Rodgersia*	Rodger's Flower
Plumbago	Lead Wort	*Rosmarinus*	Rosemary
Podocarpus	Japanese Yew	*Rubus*	Raspberry
Podocarpus	Yewpine	*Rubus*	Salmonberry
Podophyllium	May Apple	*Rubus*	Thimbleberry
Poinciana	Bird of Paradise(bush)	*Rudbeckia*	Blackeyed Susan
Polemonium	Jacob's Ladder	*Rudbeckia*	Coneflower (Orange)
Polianthes	Tuberose	*Redbeckia*	Gloriosa Daisy
Polygonatum	Solomon's Seal	*Ruellia*	Mexican Petunia
Polygonum	Fleece Flower	*Rumohra*	Leatherleaf Fern
Polygonum	Silver Lace	*Russelia*	Firecracker Palm
Polyscias	Ming Aralia	*Ruta*	Herb of Grace
Polystichum	Christmas Fern	*Sabal*	Palmetto
Pontentilla	Cinquefoil	*Sabal*	Sabal Palm
Populus	Aspen	*Sagina*	Pearlwort
Populus	Cottonwood	*Sagina*	Scotch Moss/Pearlwort
Populus	Poplar	*Saginia*	Scotch Moss (Irish)
Portulaca	Moss Rose	*Saint Paulia*	African Violet
Potentilla	Buttercup Shrub	*Salix*	Willow
Primula	Primrose	*Salvia*	Sage (Garden)
Prosopis	Mesquite	*Sambucus*	Elderberry
Prunella	Self-Heal	*Sanguinara*	Bloodroot
Prunus	Almond	*Sanguisorba*	Burnet
Prunus	Apricot	*Sansevieria*	Mother-in-Law Tongue
Prunus	Cherry	*Santolina*	Lavender Cotton
Prunus	Cherry Palm	*Sanvitalia*	Creeping Zinnia
Prunus	Choke Cherry	*Saphora*	Japanese Pagoda Tree
Prunus	English Laurel	*Sapindus*	Soapberry

SCIENTIFIC NAME – *Common Name Cross Reference*

Sapium	Chinese Tallow	*Tecoma*	Yellow Bells
Saponaria	Bouncing Bet	*Tecomaria*	Cape Honeysuckle
Saponaria	Rock Soapwart	*Ternstroemia*	Cleyera
Sarcococca	Sweet Box	*Teucrium*	Wall Germander
Saxifraga	Rockfoil	*Thailictrum*	Meadow-Rue
Saxifraga	Strawberry Begonia	*Thelypteris*	River Fern
Scabiosa	Pincushion Flower	*Thuja*	Arborvitae
Scabiosa	Star Flower	*Thuja*	Western Red Cedar
Schefflera	Umbrella Tree	*Thunbergia*	Black-eyed Susan
Schinus	California Pepper Tree	*Thymophylla*	Golden Fleece
Schizophragma	Climbing Hydrangea	*Thymus*	Mother of Thyme
Schizostylis	Kaffer Lily	*Tiarella*	Foam Flower
Sciadopitys	Umbrella Pine	*Tibouchina*	Glory Bush
Sedum	Stone Crop	*Tibouchina*	Princess Flower
Selaginella	Resurrection Plant	*Tigridia*	Mexican Shell Flower
Selaginella	Spikemoss	*Tigridia*	Tiger Flower
Sempervivum	Hens & Chickens	*Tilia*	Basswood
Senecio	Dusty Miller	*Tilia*	Linden
Senecio	Leopards Bane	*Tipuana*	Tipu Tree
Senecio	Mexican Flame Vine	*Tithonia*	Mexican Sunflower
Sequoia	Coast Redwood	*Torenia*	Wishbone Plant
Setcreasea	Purple Heart	*Trachaelospermum*	Asian Jasmine
Shepherdia	Buffalo Berry	*Trachaelospermum*	Star Jasmine
Sidalcea	Prairie Mallow	*Trachycarpus*	Windmill Palm
Silene	Catchfly	*Tradescantia*	Purple Heart
Simmondsia	Jojoba	*Tradescantia*	Spiderwort
Sinningia	Gloxinia	*Tradescantia*	Wandering Jew
Sisyrinchium	Blue-Eyed Grass	*Tragopogon*	Goat's Beard
Smilacina	Starflower	*Tricyrtis*	Toad Lily
Solandra	Cup of Gold Vine	*Trifolium*	Clover
Solidago	Goldenrod	*Trillium*	Wake Robin
Sophora	Mountain Laurel	*Tristania*	Brisbane Box Tree
Sophora	Pagoda Tree	*Tritoma*	Red Hot Poker
Sorbaria	Ash Leaf Spiraea	*Trollius*	Globe Flower
Sorbus	Mountain Ash	*Tropaeolum*	Nasturtium
Sparaxis	Harlequin Flower	*Tsuga*	Hemlock
Sparaxis	Wand Flower	*Tulippa*	Tulip
Spathiphyllum	Closet Plant	*Turnera*	Buttercup
Spigelia	Indian Pink	*Tweepia*	Southern Star
Spiraea	Bridal Wreath	*Ulmus*	Elm
Sprekella	Aztec Lily	*Umbellularia*	California Laurel
Stachys	Lamb's Ear	*Vaccinium*	Blueberry
Stismaphyllon	Butterfly Vine	*Vaccinium*	Huckleberry
Stokesia	Aster	*Vallota*	Scarborough Lily
Strelitzia	Bird of Paradise	*Vernonia*	Ironweed
Styrax	Snowbell	*Veronica*	Speedwell
Swietenia	Mahogany	*Viburnum*	Am. Cranberry Bush
Symphoricarpos	Coralberry	*Viburnum*	Arrow Wood
Symphoricarpus	Snowberry	*Viburnum*	Nannyberry
Symphytum	Russian Comfrey	*Viburnum*	Snowball
Symplocos	Sweetleaf	*Vinca*	Common Myrtle
Syngonium	Nepathytis	*Vinca*	Periwinkle
Syngonium	White Butterfly	*Viola*	Johnny Jump-up
Syringa	Lilac	*Viola*	Pansy
Syzygium	Brush Cherry	*Viola*	Violet
Tabebuia	Trumpet Tree	*Vitex*	Chaste Tree
Tagetes	Marigold	*Vitex*	Lavender Tree
Talinum	Fame Flower	*Vitis*	Grape
Tanacetum	Feverfew	*Waldsteinia*	Barren Strawberry
Tanacetum	Tansey	*Washingtonia*	California Fan Palm
Taxodium	Bald Cypress	*Washingtonia*	Mexican Fan Palm
Taxus	Yew	*Xanthoceras*	Hyacinth Shrub

Common Name Cross Reference – SCIENTIFIC NAME

Xanthoceras   Yellow Horn
Xeranthemum   Immortelle
Zantedeschia   Calla Lily
Zephyranthes   Fairy Lily
Zephyranthes   Rain Lily

COMMON NAME – *Scientific Name Cross Reference*

Common	Scientific	Common	Scientific
Aaron's Beard	*Hypercum*	Bearberry	*Arctostaphylos*
African Corn Lily	*Ixia*	Bearberry	*Cotoneaster*
African Daisy	*Osteospermum*	Beard Tongue	*Penstemon*
African Lily	*Agapanthus*	Beauty Berry	*Callicarpa*
African Violet	*Saint Paulia*	Beauty Bush	*Kolkwitzia*
Alder	*Alnus*	Bee Balm	*Monarda*
Algerian Ivy	*Hedera*	Beech	*Fagus*
Almond	*Prunus*	Beefsteak Plant	*Acalypha*
Alpine Currant	*Ribes*	Beefsteak Plant	*Perilla*
Althaea	*Hibiscus*	Bell Flower	*Campanula*
Alum Root	*Heuchera*	Bells of Ireland	*Moluccella*
Alyssum	*Lobularia*	Birch	*Betula*
Amaryllis	*Hippeastrum*	Bird of Paradise	*Stelitzia*
Amazon Lily	*Eucharis*	Bird of Paradise (Bush)	*Caesalpinia (Gilliesii)*
American Cranberry Bush	*Viburnum*	Bird of Paradise (Bush)	*Poinciana*
Amethyst Flower	*Browallia*	Bird's Nest Fern	*Nephrolepis*
Amur Corktree	*Phellodendron*	Bishop's Weed	*Aegopodium*
Angel Wing	*Jasminum*	Bittersweet	*Celastrus*
Angel's Trumpet	*Brugmansia*	Black-eyed Susan	*Thunbergia*
Angel's Trumpet	*Datura*	Blackberry Lily	*Belamcanda*
Apple	*Malus*	Blackeyed Susan	*Rudbeckia*
Apricot	*Prunus*	Blackgum	*Nyssa*
Aralia	*Fatsia*	Bladder Pod	*Alyssoides*
Arborvitae	*Thuja*	Blanket Flower	*Gaillardia*
Areca Palm	*Chrysalidocarpus*	Blazing Star	*Liatris*
Arizona Cypress	*Cupressus*	Bleeding Heart	*Corydalis*
Arrow Wood	*Viburnum*	Bleeding Heart	*Dicentra*
Ash	*Fraxinus*	Bleeding Heart Glorybower	*Clerodendrum*
Ash Leaf Spiraea	*Sorbaria*	Blood Flower	*Asclepias*
Asian Jasmine	*Trachaelospermum*	Blood Lily	*Haemanthus*
Aspen	*Populus*	Bloodroot	*Sanguinara*
Aster	*Stokesia*	Blue Bonnet	*Lupinus*
Aster (Stokes)	*Callistephus*	Blue Bush Daisy	*Felicia*
Autumn Olive	*Elaeagnus*	Blue Daze	*Evolvulus*
Avens	*Geum*	Blue Mist	*Caryopteris*
Avocado	*Persea*	Blue Star	*Amsonia*
Azalea	*Rhodedendron*	Blue Star Creeper	*Laurentia*
Aztec Lily	*Sprekella*	Blue-Eyed Grass	*Sisyrinchium*
Baby's Breath	*Gypsophila*	Bluebeard	*Caryopteris*
Bachelor Buttons	*Centaurea*	Blueberry	*Vaccinium*
Bachelor Buttons	*Gomphrena*	Bog Rosemary	*Andromeda*
Bald Cypress	*Taxodium*	Boston Fern	*Nephrolepis*
Bali High	*Dieffenbachia*	Boston Ivy	*Parthenocissus*
Balloon Flower	*Platycodon*	Botanical Wonder	*Fatshedera*
Balloon Vine	*Cardiospermum*	Bottle Tree	*Brachychiton*
Bamboo	*Bambusa*	Bottlebrush	*Callistemon*
Bamboo (dwarf)	*Pleioblastuis*	Bouncing Bet	*Saponaria*
Bamboo Palm	*Chamaedorea*	Bower Vine	*Pandorea*
Banana	*Musa*	Box Elder	*Acer*
Baneberry	*Actaea*	Boxwood	*Buxus*
Banyan	*Ficus*	Bracken Fern	*Pteridium*
Barbara's Buttons	*Marshaccia*	Bridal Wreath	*Spiraea*
Barberry	*Berberis*	Brisbane Box Tree	*Tristania*
Barren Strawberry	*Waldsteinia*	Broom	*Cytisus*
Barrenwort	*Epimedium*	Broom	*Genista*
Basket of Gold	*Aurinia*	Broom Tee Tree	*Leptospermum*
Basswood	*Tilia*	Brush Cherry	*Syzygium*
Bay Laurel (Sweet Bay)	*Laurus*	Buckeye	*Aesculus*
Bayberry	*Myrica*	Buckthorn	*Rhamnus*
Bear's Breech	*Acanthus*		

Scientific Name Cross Reference – COMMON NAME

Common	Scientific
Buffalo Berry	Shepherdia
Bugbane	Cimicifuga
Bugle Flower	Ajuga
Bugleweed	Ajuga
Bunchberry	Cornus
Burnet	Sanguisorba
Burning Bush	Euonymus
Bush Honeysuckle	Diervilla
Buttercup	Ranunculus
Buttercup	Turnera
Buttercup Shrub	Potentilla
Butterfly Bush	Buddleia
Butterfly Vine	Stismaphyllon
Butterfly Weed	Asclepias
Buttonbush	Cephalanthus
Cajeput Tree	Melaleuca
Calico Plant	Alternanthera
California Fan Palm	Washingtonia
California Laurel	Umbellularia
California Pepper Tree	Schinus
California Poppy	Eschscholzia
Calla Lily	Zantedeschia
Camass	Camassia
Candy Lily	Pardancanda
Candytuft	Iberis
Canterberry Bells	Campanula
Cape Honeysuckle	Tecomaria
Cape Iris	Homeria
Cape Jasmine	Gardenia
Cape Marigold	Dimorphotheca
Cape Weed	Arctotheca
Caramel Creeper	Ceanothus
Cardinal Flower	Lobelia
Carnation	Dianthus
Carob	Ceratonia
Carolina Allspice	Calycanthus
Caroline Jasmine	Gelsemium
Carpet Bugle	Ajuga
Carrot Wood	Cupaniopsis
Cast-Iron Plant	Aspidistra
Catchfly	Silene
Catnip	Nepeta
Cedar	Cedrus
Cedar	Chamaecyparis
Cedar (Red)	Juniperus
Cedar of Lebanon	Cedrus
Century Plant	Agave
Chamomile	Matricaria
Chaste Tree	Vitex
Chenille Plant	Acalypha
Cherry	Prunus
Cherry Brush	Eugenia
Cherry Palm	Prunus
Chestnut	Castanea
Chilean Jasmine	Mandevilla
China Blue Vine	Holboellia
China-Berry	Mella
Chinese Bell Flower	Abutilon
Chinese Evergreen	Aglaonema
Chinese Fan Palm	Livistona
Chinese Gooseberry	Actinidia
Chinese Ground Orchid	Bletilla
Chinese Pistache	Pistacia
Chinese Quince	Pseudocydonia
Chinese Tallow	Sapium
Choke Cherry	Prunus
Chokeberry	Aronia
Christmas Fern	Polystichum
Christmas Rose	Helleborus
Chrysanthemum	Dendranthema
Cinnamon Fern	Osmunda
Cinquefoil	Pontentilla
Cleyera	Ternstroemia
Climbing Hydrangea	Schizophragma
Climbing Lily	Gloriosa
Closet Plant	Spathiphyllum
Clover	Trifolium
Clump Bamboo	Fargesia
Coast Redwood	Sequoia
Cobra Lily	Arisaemia
Cockscomb	Celosia
Coffee Tree	Coffea
Coffeeberry	Rhamnus
Columbine	Aquilegia
Common Myrtle	Vinca
Coneflower	Echinacea
Coneflower	Ratibida
Coneflower (Orange)	Rudbeckia
Copper Plant	Acalypha
Copperleaf	Acalypha
Coral Beauty	Contoneaster
Coral Bells	Heuchera
Coral Vine	Antigonon
Coralberry	Symphoricarpos
Corn Lily	Ixia
Corn Plant	Dracaena
Cornbells	Ixia
Cornelian Cherry	Cornus
Coromandel	Asystasia
Cottonwood	Populus
Coyote Bush	Baccharis
Crabapple	Malus
Cracena	Cordyline
Cranberry	Cotoneaster
Cranesbill	Geranium
Crape Myrtle	Lagerstroemia
Creeping Charlie	Pilea
Creeping Jenny	Lysimachia
Creeping Zinnia	Sanvitalia
Crimson Pygmy	Berberis
Croton	Codiaeum
Crownvetch	Coronilla
Cuban Laurel	Ficus
Cup of Gold Vine	Solandra
Cupflower	Nierembergia
Cupid's Dart	Catananche
Cushion Mums	Chrysanthemum
Daffodil	Narcissus
Dahlberg Daisy	Dyssodia
Daisy Bush	Felicia
Daisy Fleabane	Erigeron
Dancing Girls	Globba

COMMON NAME – *Scientific Name Cross Reference*

Common	Scientific
Date Palm	*Phoenix*
Day Lily	*Hemerocallis*
Dead Nettle	*Lamium*
Desert Rose	*Adenium*
Desert Willow	*Chilopsis*
Dogtooth Violet	*Erythronium*
Dogwood	*Cornus*
Douglas Fir	*Pseudotsuga*
Dove Tree	*Davidia*
Dragonhead	*Dracephalium*
Dragonhead	*Physostegia*
Dumb Cane	*Dieffenbachia*
Dusty Miller	*Centaurea*
Dusty Miller	*Senecio*
Dutchman's Breeches	*Dicentra*
Dwarf Mountain Lover	*Paxistima*
Dwarf Plumbago	*Ceratostigma*
Dyers Wood	*Isatis*
Edelweiss	*Leontopodium*
Egret Lily	*Habenaria*
Elderberry	*Sambucus*
Elephant Ears	*Colocasia*
Elephant's Ear	*Alocasia*
Elm	*Ulmus*
Empress Tree	*Paulownia*
English Daisy	*Bellis*
English Ivy	*Hedera*
English Laurel	*Prunus*
Eulalia Grass	*Miscanthus*
European Ginger	*Asarum*
European Liver-Leaf	*Hepatica*
Evening Primrose	*Oenothera*
Fairy Lily	*Zephyranthes*
False Aralia	*Dizgsotheca*
False Cypress	*Chamaecyparis*
False Dragonhead	*Physostegia*
False Heather	*Cuphea*
False Indigo	*Amorpha*
False Lilly of the Valley	*Maianthemum*
False Spirea	*Astibe*
False Sunflower	*Heliopsis*
Fame Flower	*Talinum*
Fan Palm	*Chamaerop*
Fescue	*Festuca*
Feverfew	*Chrysanthemum*
Feverfew	*Matricaria*
Feverfew	*Tanacetum*
Fiesta Daisy	*Gaillardia*
Fiesta Daisy	*Gallardia*
Fig	*Ficus*
Filbert	*Corylus*
Fir	*Abies*
Firebush	*Hamelia*
Firecracker Palm	*Caryota*
Firecracker Palm	*Russelia*
Firetails	*Acalypha*
Firethorn	*Pyracantha*
Fishtail Palm	*Caryota*
Flame of the Woods	*Ixora*
Flax	*Linum*
Fleece Flower	*Polygonum*
Floss Flower	*Ageratum*
Flowering Kale	*Brassicaceae*
Flowering Maple	*Abutilon*
Flowering Tobacco	*Nicotiana*
Foam Flower	*Tiarella*
Fool's Huckleberry	*Menziesia*
Forget-Me-Not	*Brunnera*
Forget-Me-Not	*Myosotis*
Forsythia (white)	*Abeliophyllum*
Fortnight Lily	*Morea*
Fountain Grass	*Pennisetum*
Four-O-Clock	*Mirabilis*
Foxglove	*Digitalis*
Foxtail Lily	*Eremurus*
Fringe Flower	*Loropetalum*
Fringe Tree	*Chionanthus*
Fuchsia (Australian)	*Correa*
Garden Gloxinia	*Incarvillae*
Gardeners Garters	*Phalaris*
Gas Plant	*Dictamnus*
Gayfeather	*Liatris*
Geranium	*Pelargonium*
Giant Snail Vine	*Phaseolius*
Ginger	*Asarum*
Ginger	*Hedychium*
Globe Amaranth	*Gomphrena*
Globe Flower	*Kerria*
Globe Flower	*Trollius*
Globe Thistle	*Echinops*
Gloriosa Daisy	*Rudbeckia*
Glory Bush	*Tibouchina*
Glorybower	*Clerodendrum*
Gloxinia	*Sinningia*
Goat's Beard	*Aruncus*
Goat's Beard	*Astibe*
Goat's Beard	*Tragopogon*
Gold Dust Plant	*Aucuba*
Golden Aster	*Chrysopsis*
Golden Bells	*Forsythia*
Golden Chain	*Labumum*
Golden Daisy	*Coreopsis*
Golden Fleece	*Thymophylla*
Golden Globe	*Lysimachia*
Golden Larch	*Pseudolarix*
Goldenrain Tree	*Koelreuteria*
Goldenrod	*Solidago*
Goldenstar	*Chrysogonum*
Grape	*Vitis*
Grape Hyacinth	*Muscari*
Grape Ivy	*Cissus*
Green Gold	*Chrysogonum*
Ground Ivy	*Glechoma*
Guava	*Feijoa*
Guinea Gold Vine	*Hibbertia*
Hackberry	*Celtis*
Hardy Gloxinia	*Incarvillea*
Harebell	*Campanula*
Harlequin Flower	*Sparaxis*
Hawaiin Ti	*Cordyline*
Hawaiian Tree Fern	*Cibotium*
Hawthorn	*Crataegus*
Hay Sentel Fern	*Dennstaedtia*
Hazelnut	*Corylus*

Scientific Name Cross Reference – **COMMON NAME**

Common	Scientific
Heartpea	Cardiospermum
Heath	Erica
Heath (Irish)	Daboecia
Heather	Calluna
Heather	Cassiope
Heather	Erica
Heavenly Bamboo	Nandina
Hellebore	Helleborus
Hemlock	Tsuga
Hens & Chickens	Sempervivum
Herb of Grace	Ruta
Hickory	Carya
Hidden Lily	Curcuma
Himalayan Blue Poppy	Meconopsis
Hinoki Cypress	Chamaecyparis
Holly	Ilex
Holly Fern	Arachniodes
Holly Fern	Cyrtomium
Holly Olive	Osmanthus
Hollyhock	Alcea
Honesty Plant	Lunaria
Honey Locust	Gleditsia
Honeysuckle	Lonicera
Hophornbean	Ostrya
Hoptree	Ptelea
Hornbean	Carpinus
Horse Chestnut	Aesculus
Huckleberry	Vaccinium
Hyacinth Shrub	Xanthoceras
Hyssop	Agastiache
Ice Plant	Mesembryanthemum
Iceplant	Carpobrotus (perennials)
Iceplant	Cryophytum (Annual)
Iceplant (trailing)	Drosanthemum
Iceplant (white/hardy)	Delosperma
Immortelle	Xeranthemum
Incense Cedar	Calocedrus
Incense Cedar	Libocedrus
Indian Blanket	Gallardia
Indian Hawthorn	Raphioepis
Indian Laurel	Ficus
Indian Pink	Spigelia
Indian Plum	Oemleria
Indigo	Bapitsia
Inkberry	Ilex
Ironweed	Vernonia
Italian Cypress	Cupressus
Jack in the Pulpit	Arisaemia
Jacob's Ladder	Polemonium
Jade Plant (Tree)	Crassula
Japanese Andromeda	Pieris
Japanese Angelica Tree	Aralia
Japanese Cedar	Cryptomeria
Japanese Pagoda Tree	Saphora
Japanese Spurge	Pachysandra
Japanese Yew	Podocarpus
Job's Tears	Coix
Joe Pye Weed	Eupatorium
Johnny Jump-Up	Viola
Jojoba	Simmondsia
Joseph's Coat	Alternanthera
Juneberry	Amalanchier
Juniper	Juniperus
Jupiter's Beard	Centranthus
Kafer Lily	Clivia
Kaffer Lily	Schizostylis
Kangaroo Ivy	Cissus
Katsura Tree	Cercidiphyllum
Kentia Palm	Howea
Kentucky Coffee Tree	Gymnocladus
Keys of Heaven	Centranthus
Kiwi	Actinidia
Lady Bells	Adenophora
Lady Fern	Athyrium
Lady Palm	Rhapis
Lady's Mantle	Alchemilla
Lamb's Ear	Stachys
Lapa Lapa	Cheirodendron
Larch	Larix
Larkspur	Consolida
Larkspur	Delphinum
Laurel Cherry	Prunus
Lavender	Lavendula
Lavender Cotton	Santolina
Lavender Tree	Vitex
Lead Wort	Plumbago
Leaf Ivy	Cissus
Leatherleaf Fern	Rumohra
Lemon Verbena	Aloysia
Lenten Rose	Helleborus
Leopards Bane	Doronicum
Leopards Bane	Senecio
Leyland Cypress	Cupressocyparis
Lilac	Syringa
Lilac Hibiscus	Alyogyne
Lily	Lillum
Lily (Easter)	Longiflorum
Lily of the Nile	Agapanthus
Lily of the Valley	Convallaria
Lily of the Valley (shrub)	Pieris
Lilyturf	Liriope
Linden	Tilia
Lipstick Plant	Aeshcynanthus
Livingstone Daisy	Mesembryanthemum
Locust (Black)	Robinia
Loosestrife	Lythrum
Loquat	Eribobotrya
Lotus	Nelumbo
Love in a Mist	Nigella
Lungwort	Pulmonania
Madrone	Arbutus
Magic Lily	Lycoris
Mahogany	Swietenia
Maidenhair Tree	Ginkgo
Maidenhead Fern	Adiantum
Majestic Palm	Ravenea
Mallow	Malva
Maltese Cross	Lychnis
Manzanita	Arctostaphylos
Maple	Acer
Marginata	Dracaena
Marguerite	Chrysanthemum

COMMON NAME – *Scientific Name Cross Reference*

Common Name	Scientific Name
Margerite (Golden)	*Anthemis*
Marigold	*Tagetes*
Marsh Marigold	*Caltha*
Masterwort	*Astrantia*
May Apple	*Podophyllium*
Meadow Beauty	*Rhexia*
Meadow Sweet	*Filipendula*
Meadow-Rue	*Thailictrum*
Mesquite	*Prosopis*
Metallic Plant	*Hemi*
Mexican Creeper	*Antigonon*
Mexican Fan Palm	*Washingtonia*
Mexican Flame Vine	*Senecio*
Mexican Heather	*Cuphea*
Mexican Honeysuckle	*Justicia*
Mexican Orange	*Choisya*
Mexican Petunia	*Ruellia*
Mexican Shell Flower	*Tigridia*
Mexican Sunflower	*Tithonia*
Michaelmas Daisy	*Aster*
Mickey Mouse Bush	*Ochna*
Milfoil	*Achillea*
Milkweed	*Asclepias*
Mimosa	*Albizia*
Ming Aralia	*Polyscias*
Mirror Plant	*Caprosma*
Mist Flower	*Eupatorium*
Mock Orange	*Philadelphus*
Mondo Grass	*Ophiopogon*
Money Plant	*Lunaria*
Moneywort	*Lysimachia*
Monkey Flower	*Mimulus*
Monkey Puzzle Tree	*Araucaria*
Monkshood	*Aconitum*
Moon Flower	*Calonyction*
Morninglory	*Impomea*
Moss Pink	*Phlox*
Moss Rose	*Portulaca*
Moss Sandwort	*Arenaria*
Mother of Thyme	*Thymus*
Mother-in-Law Tongue	*Sansevieria*
Motherwort	*Leonurus*
Mountain Ash	*Pyrus*
Mountain Ash	*Sorbus*
Mountain Holly	*Nemopanthus*
Mountain Laurel	*Kalmia*
Mountain Laurel	*Sophora*
Mountain Mahogany	*Cercocarpus*
Mt. Atlas Daisy	*Anacyclus*
Mulberry	*Morus*
Musk Mallow	*Abelmoschus*
Myrtle	*Myrtus*
Nannyberry	*Viburnum*
Nasturtium	*Tropaeolum*
Natal Plum	*Carissa*
Nepathytis	*Syngonium*
Nerve Plant	*Fittonia*
New Zealand Flax	*Phormium*
Night-blooming Jasmine	*Cestrum*
Ninebark	*Physocarpus*
Norfolk Island Pine	*Araucaria*
Oak	*Quercus*
Obedient Plant	*Physostegia*
Ocean Spray	*Holodiscus*
October Plant Perilla	*Hylatelephium*
Oleander	*Nerium*
Olive	*Olea*
Orange Jessamine	*Murraya*
Orchid	*Dendrobium*
Oregon Boxwood	*Paxistima*
Oregon Grape	*Mahonia*
Oriental Arborvitae	*Platycladus*
Osage Orange	*Maclura*
Ostrich Fern	*Pteretis*
Ox-Eye Daisy	*Chrysanthemum*
Ox-Eye Daisy	*Leucanthemum*
Oyster Plant	*Rhoeo*
Pagoda Tree	*Sophora*
Painted Daisy	*Chrysanthemum*
Palm	*Chamaedorea*
Palmetto	*Sabal*
Palo Verde	*Cercidium*
Palo Verde	*Parkensonia*
Pampas Grass	*Cortaderia*
Panda Plant	*Kalanchoe*
Pandora Vine	*Pandorea*
Pansy	*Viola*
Papaya	*Carcia*
Parlor Palm	*Chamaedorea*
Parrot's Beak	*Lotus*
Pasque Flower	*Anemone*
Pasque Flower	*Pulsatilla*
Passion Vine	*Gynura*
Passionflower	*Passiflora*
Paw Paw	*Asimina*
Peach	*Prunus*
Peacock Plant	*Calathea*
Pear	*Pyrus*
Pearl Bush	*Exochorda*
Pearlwort	*Sagina*
Peashrub	*Caragana*
Pecan	*Carya*
Peony	*Paeonia*
Pepper	*Capsicum*
Peppermint	*Mentha*
Periwinkle	*Vinca*
Periwinkle (Madagascar)	*Catharanthus*
Periwinkle (Vinca)	*Catharanthus*
Persian Cornflower	*Centaurea*
Persian Violet	*Cyclamen*
Persimmon	*Diospyros*
Peruvian Daffodil	*Ismene*
Peruvian Lily	*Alstroemeria*
Philodendron	*Monstera*
Pimpernel	*Anagallis*
Pincushion Flower	*Scabiosa*
Pindo Palm	*Butia*
Pine	*Pinus*
Pineapple Lily	*Eucomis*
Pink	*Dianthus*
Pink Lady	*Raphiolepis*
Plane Tree	*Platanus*

Scientific Name Cross Reference – COMMON NAME

Common Name	Scientific Name
Plantain Lily	Hosta
Plum	Prunus
Plum Yew	Cephalotaxus
Plubago	Ceratostigma
Plume Poppy	Macleaya
Poker Plant	Kniphofia
Polka Dot Plant	Hypoestes
Pomegranate	Punica
Ponytail Palm	Beaucarnea
Poplar	Populus
Poppy	Papaveer
Poppy Mallow	Callirhoe
Porcelain Vine	Ampelopsis
Pot Marigold	Calendula
Pothos Ivy	Epipremnum
Prairie Mallow	Sidalcea
Prayer Plant	Maranta
Primrose	Primula
Primrose Jasmine	Jasminum
Princess Flower	Tibouchina
Privet	Ligustrum
Purple Coneflower	Echinacea
Purple Fringe	Cotinus
Purple Heart	Setcreasea
Purple Hopseed Bush	Tradescantia
Purple Leaf	Dodonaea
Purple Leaf Sand Cherry	Prunus
Queen Anne's Lace	Daucus
Queen of the Prairie	Filipendula
Queen of the Prairies	Filipendula
Queen Palm	Arecastrum
Queens Wreath	Antigonon
Quince	Chaenomeles
Rain Lily	Zephyranthes
Raspberry	Rubus
Raspberry Ice	Bougainvillea
Red Bay	Persea
Red Fountain Grass	Pennisetum
Red Gum	Eucalyptus
Red Hot Poker	Tritoma
Red Iron Bark	Eucalyptus
Redbud	Cercis
Redroot Jersey Tea	Ceanothus
Redwood	Metasequoia
Reflexa	Dracaena
Resurrection Plant	Selaginella
River Fern	Thelypteris
Robira	Pittsporium
Rock Cress	Arabis
Rock Cress	Aubrieta
Rock Rose	Cistus
Rock Rose	Helianthemum
Rock Rose	Pavonia
Rock Soapwart	Saponaria
Rockfoil	Saxifraga
Rodger's Flower	Rodgersia
Rose Champion	Lychnis
Rose Mallow	Hibiscus
Rose of Sharon	Althea
Rose of Sharon	Hibiscus
Rosemary	Rosmarinus
Royal Fern	Osmunda
Rubber Plant (tree)	Ficus
Rubber Vine	Cryptostesia
Rupturewort	Herniaria
Russian Comfrey	Symphytum
Russian Olive	Elaeagnus
Russian Sage	Perovskia
Sabal Palm	Sabal
Sage (Garden)	Salvia
Sago Palm	Cycas
Salmonberry	Rubus
Sand Verbena	Abronia
Scarborough Lily	Vallota
Schefflera	Brassaia
Scotch Moss (Irish)	Saginia
Scotch Moss/Pearlwort	Saginia
Scotchbroom	Cytisus
Screw Pine	Pandanus
Sea Bushthorn	Hippophae
Sea Daffodil	Pancratium
Sea Holly	Erymgium
Sea Lavendar	Limonium
Sea Pinks	Armeria
Sedge	Carex
Self-Heal	Prunella
Selloum	Philodendron
Sensitive Fern	Onoclea
Sensitive Plant	Mimosa
Serviceberry	Amelanchier
Shadbush	Amalanchier
Shasta Daisy	Chrysanthemum
Shasta Daisy	Leucanthemum
Shoo-Fly Plant	Nicandra
Shooting Star	Dodecatheon
Shooting Star	Pseuderanthemum
Shrimp Plant	Justicia
Siberian Pea Tree	Caragana
Silk Tassel	Garrya
Silktree	Albizia
Silver Frost	Lamiastrum
Silver King	Euonymus
Silver Lace	Polygonum
Silverbells	Halesia
Silverberry	Elaeagnus
Sky Flower	Duranta
Smokebush	Cotinus
Smoketree	Cotinus
Snake Root	Cimicifuga
Snakehead	Chelone
Snapdragon	Antirrhimum
Snow in Summer	Cerastium
Snowball	Viburnum
Snowbell	Styrax
Snowberry	Symphoricarpus
Snowflake	Leucojum
Soapberry	Sapindus
Solomon's Seal	Polygonatum
Sour Gum	Nyssa
Sourwood	Oxydendrum
Southern Star	Tweepia
Spanish Bluebell	Endymion

119

COMMON NAME – *Scientific Name Cross Reference*

Common Name	Scientific Name	Common Name	Scientific Name
Speedwell	*Veronica*	Tiger Flower	*Tigridia*
Speenwort Fern	*Athyrium*	Tiger Lily	*Lilium*
Spicebush	*Illicium*	Tipu Tree	*Tipuana*
Spicebush	*Lindera*	Toad Lily	*Tricyrtis*
Spider Lily	*Lycoris*	Toyon	*Leptospermum*
Spider Plant	*Chiorophytum*	Transvaal Daisy	*Gerbera*
Spider Plant	*Cleome*	Treasue Flower	*Ganzania*
Spiderwort	*Tradescantia*	Tree Mallow	*Lavatera*
Spikemoss	*Selaginella*	Triangle Palm	*Neodypsis*
Spindle Tree	*Euonymus*	Trumpet Creeper	*Campsis*
Sprengeri Fern	*Asparagus*	Trumpet Crossvine	*Bignonia*
Spruce	*Picea*	Trumpet Tree	*Tabebuia*
Spur Flower	*Plectranthus*	Trumpet Vine	*Pandorea*
Spurge	*Euphorbia*	Tuberose	*Polianthes*
Squaw Carpet	*Ceanothus*	Tulip	*Tulippa*
St. Johns Wort	*Hypericum*	Tulip Tree	*Linodendron*
Staghorn Fern	*Platycerium*	Tunic Flower	*Petrorhagia*
Star Flower	*Ipiteion*	Tupelo	*Nyssa*
Star Flower	*Omithogalum*	Turk's Cap	*Malvauiscus*
Star Flower	*Scabiosa*	Turtlehead	*Chelone*
Star Jasmine	*Trachaelospermum*	Twin Flower	*Linnaea*
Star of Bethlehem	*Omithogalum*	Twinspur	*Diascia*
Star of India	*Achimenes*	Umbrella Pine	*Sciadopitys*
Star Plant	*Grewia*	Umbrella Plant	*Cyperus*
Starflower	*Smilacina*	Umbrella Plant	*Darmera*
Statice	*Limonium*	Umbrella Tree	*Brassaia*
Stock	*Matthiola*	Umbrella Tree	*Schefflera*
Stone Crop	*Sedum*	Valerian	*Centranthus*
Straw Flower	*Helichrysan*	Veronica	*Hebe*
Strawberry	*Fragaria*	Violet	*Viola*
Strawberry Begonia	*Saxifraga*	Virginia Creeper	*Parthenocissus*
Strawberry Tree	*Arbutus*	Wake Robin	*Trillium*
Sumac	*Rhus*	Walking Stick	*Aralia*
Summersweet	*Clethra*	Wall Flower	*Cheiranthus*
Sun Rose	*Helianthemum*	Wall Germander	*Teucrium*
Sundrops	*Oenothera*	Wallflower	*Cheiranthus*
Sunflower	*Helianthus*	Wallflower	*Erysimum*
Swan River Daisy	*Brachycome*	Walnut	*Juglans*
Swedish Ivy	*Plectranthus*	Wand Flower	*Sparaxis*
Sweet Box	*Sarcococca*	Wandering Jew	*Tradescantia*
Sweet Grass (Flag)	*Acorus*	Water Elm	*Planera*
Sweet Pea	*Lathyrus*	Water Lily	*Nymphaes*
Sweet Pepperbush	*Clethra*	Wattle	*Acacia*
Sweet William	*Dianthus*	Wax Myrtle	*Myrica*
Sweet Woodruff	*Galium*	Wax Plant	*Hoya*
Sweetgum	*Liquidambar*	Weeping Fig	*Ficus*
Sweetleaf	*Symplocos*	West Indian Holly	*Leea*
Sweetspire	*Itea*	Western Red Cedar	*Thuja*
Sycamore	*Platanus*	White Butterfly	*Syngonium*
Table Fern	*Pteretis*	White Sage	*Artemisia*
Tallhedge	*Rhamnus*	Wild Lilac	*Ceanothus*
Tansey	*Tanacetum*	Willow	*Salix*
Tea Tree	*Laptospermum*	Wind Flower	*Anemone*
Terrestrial Orchid	*Pleione*	Windmill Palm	*Trachycarpus*
Texas Bluebell	*Eustoma*	Winter Creeper	*Eunoymus*
Texas Sage	*Ceniza*	Winterberry	*Ilex*
Texas Sage	*Leucophyllum*	Wintergreen	*Gaultheria*
Thimbleberry	*Rubus*	Winterhazel	*Corylopsis*
Thorn Apple	*Datura*	Wintersweet	*Chimonamthus*
Thrift	*Armeria*	Wishbone Plant	*Torenia*
Thrift	*Phlox*	Witch Hazel	*Hamamelis*
Tickseed	*Coreopsis*	Woadwaxen	*Genista*

Scientific Name Cross Reference – **COMMON NAME**

Wood Fern	*Dryopteris*
Wood Sorrel	*Oxalis*
Woodbine	*Lonicera*
Woody Sunflower	*Eriophyllum*
Wormwood	*Artemisia*
Yarrow	*Achillea*
Yaupon	*Ilex*
Yellow Archangel	*Lamium*
Yellow Bells	*Tecoma*
Yellow Horn	*Xanthoceras*
Yellow Poplar	*Liriodendron*
Yellow Wood	*Cladrastis*
Yew	*Taxus*
Yewpine	*Podocarpus*
Yucca	*Hesperaloe*
Zebra Grass	*Miscanthus*
Zebra Plant	*Aphelandra*